THE BOMB

ALSO BY STEPHEN M. YOUNGER

Endangered Species:
How We Can Avoid Mass Destruction
and Build a Lasting Peace

An Imprint of HarperCollins*Publishers*

A NEW HISTORY

Stephen M. Younger

A hardcover edition of this book was published in 2009 by Ecco, an imprint of HarperCollins Publishers.

HarperCollins books may be purchased for educational, business, or sales promotional use. For information, please write: Special Markets Department, HarperCollins Publishers, 10 East 53rd Street, New York, NY 10022.

First Ecco paperback edition published 2010.

Designed by Cassandra J. Pappas

Library of Congress Cataloging-in-Publication Data is available upon request.

ISBN: 978-0-06-153720-2

10 11 12 13 14 ID/RRD 10 9 8 7 6 5 4 3 2 1

Contents

Illustrations

Acknowledgments

It is a pleasure to acknowledge many helpful conversations with James Mercer-Smith and David Sharp, both of Los Alamos National Laboratory. Hanaa Benhalim of the Defense Threat Reduction Agency provided invaluable assistance in locating photographs of U.S. weapons systems. Kathy Vinson, of the Defense Visual Information Center, was very helpful in locating images of missiles, aircraft, and submarines. Mim John offered thoughtful comments on alternative views along with many specific suggestions for improvement. I would like to recognize admirals Richard Mies and James Ellis, both former commanders of United States Strategic Command, for their leadership and vision in transforming strategic thinking. My editors at Ecco, Emily Takoudes and Greg Mortimer, and my agents, Don Lamm and Christy Fletcher, provided sage advice throughout this project. Special thanks go to my wife, Mari, for her support and critical reading of the manuscript.

AUTHOR'S NOTE

I regret that I am unable to provide a reference list to direct the interested reader to more detailed information on nuclear weapons and their history. Security regulations prohibit me from commenting on or even referring to any publication that has received a "no comment" ruling from the United States government. My access to classified information could, by itself, inadvertently lend credibility to any cited source. Therefore, in the interests of fairness and security, I have chosen to exclude notes and references from this book.

THE BOMB

Introduction

Why Nuclear Weapons in the Twenty-first Century?

They had been worried by thunderstorms. In the pre-dawn darkness, men gathered in a makeshift desert camp, the lucky ones busying themselves with dials and gauges, cables and checklists. The rest paced back and forth, smoked, and anxiously watched the clock. There was nervous conversation: What if it didn't work? What if it did? So much had been invested in this one event, precious resources in a time of war, all based on a promise, a theory—and, some might say, scientific arrogance.

At the appointed time a brilliant flash lit up the sky, a flash brighter than a thousand suns, a flash that penetrated the thick welders' goggles worn by the observers, making them wince. It was a silent flash, the light moving so much faster than the shock wave that followed it. Seconds later came the bang, and then the steady roar of the atmosphere being displaced by a heat never before known on earth. A titanic mushroom cloud rose from the desert floor, a boiling mass of dust and gas. Purple lightning flashed across it. Those who saw it stood in stunned silence, aware that this was something new, a turning point, a single explosion in history that would reverberate through years, decades, perhaps even centuries. The men standing in the New Mexico desert on July 16, 1945, were its creators, those who had pulled the stopper on the bottle that contained the nuclear genie. Their wish was for their creation to end a war, but as with all genies, there were unintended consequences.

At the start of the twenty-first century, the United States stands at a position of unprecedented military superiority. No other country, or even most combinations of countries, could hope to defeat America in a conventional battle on land, at sea, in the air, or in space. Our technological prowess has enabled us to field weapons that can literally run rings around our competitors, giving us a superior fighting capability with fewer tanks, planes, and ships than any of our potential adversaries. More than just technology, we have military personnel who would have been the envy of any general or admiral of the past—well educated, highly motivated, and dedicated to their jobs. Finally, the United States has demonstrated the will to use military force in the pursuit of its national objectives. America is no shrinking violet that might tempt an aggressive challenger—everyone understands that America is *the* force to be reckoned with in the world.

The only real threat to U.S. military forces comes from nuclear, chemical, or biological weapons. Of the three, only nuclear weapons could inflict tactical defeat upon our forces. American troops are trained to fight and to win in environments contaminated with chemical or biological agents, but when a nuclear explosion occurs, there is nothing that can be done except to endure the blast, heat, and radiation, and hope that you are far enough away to survive. Protective suits and special vehicles enable soldiers to continue to function and advance even in thick clouds of phosgene, chlorine, and anthrax, but neither armor nor protective gear can protect against temperatures hot enough to melt steel and pressures great enough to crush a tank like a tin can. The United States has become the ultimate combat force on the planet—only the ultimate weapon can threaten that force.

But if nuclear weapons are the only real threat to our high-tech soldiers, sailors, and airmen, why don't we have a concerted policy to eliminate them from the planet? Wouldn't it make sense to just get rid of these uniquely threatening weapons so that the United States would reign supreme on any future battlefield? Why would we even *consider* the construction of new weapons, an action that might start another nuclear arms race that would only serve to put America at *greater* risk?

Conversely, when has history given any nation a permanent lock on power, one that will never be challenged by an adversary who would secretly violate international agreements to gain a strategic advantage? In the arms race preceding the Second World War, both Germany and Japan ignored treaties intended to cap the destructive capabilities of weapons. Each developed super-battleships, advanced aircraft, and more powerful artillery while their future adversaries stuck to their promises. Now, in another period of uncertainty and change, we are faced with the decision of what to do with the most powerful weapons ever created, the nuclear arsenals of the eight or ten nations in the world that have crossed a fundamental threshold in the ability to destroy.

Nuclear weapons, more so than any weapons that preceded them, touch not just the warrior but the citizen. As weapons of mass destruction, they protect—and threaten—the very foundations of civilization. The time has come to bring discussions of nuclear policy out of the cloistered enclave of defense strategists to engage a much wider audience. Nuclear weapons affect all of us, and all of us should have the opportunity to discuss their future.

NUCLEAR WEAPONS WERE the icons of the Cold War. Nothing so graphically illustrated the seriousness of the superpower standoff as did images of nuclear-armed bombers rising off the ends of runways and missiles launched from underground silos. Humankind has always had the capability to inflict destruction; with the advent of the nuclear age the character of that destruction changed in a fundamental way. A full-scale nuclear exchange between the United States and the Soviet Union would not only have killed hundreds of millions of people, it would have destroyed cities and industrial plants, and done enormous damage to the global environment. It would have dealt a staggering and perhaps even a fatal blow to the development of human civilization. It is astonishing that, even though the fear of Armageddon was palpable at times, that fear did not prevent the United States and the Soviet Union from building tens of thousands of nuclear weapons or from continuing to explore advanced designs that put ever more destruction into ever smaller packages. Not only did we walk up to the brink of worldwide destruction, we walked *along* that brink for decades.

Numerous attempts were made to control the environmental damage done by the testing of nuclear weapons and to slow the inexorable rise of their destructive power. Treaties were put into place to prohibit nuclear explosions in the atmosphere, in the ocean, or in space, and to limit the explosive force of nuclear tests conducted underground. Strategic arms treaties placed caps on the number of missiles, bombers, and submarines that a nation could have, caps that could be verified by regular inspections. The Nuclear Nonproliferation Treaty of 1970 tried to halt the spread of nuclear weapons beyond the five nations that admitted to having them (the United States, the Soviet Union, Great Britain, France, and China) by promising to provide peaceful nuclear technology to countries that agreed not to develop weapons. A Comprehensive Test Ban Treaty, forbidding any type of nuclear explosion, was signed by many countries in 1996. The United States played a central role in the development and implementation of all these treaties, and although it did not ratify the Comprehensive Test Ban Treaty, it has abided by its provisions and has not conducted a nuclear test since 1992. While recent years have seen nuclear tests by India, Pakistan, and North Korea, and development programs in Iran and elsewhere, the number of countries that chose the nuclear path is much smaller than was projected during the 1960s.

Looking backward from a world focused on the dangers of transnational terrorism, one in which stories of the proliferation of nuclear weapons technology are on the front page almost every day, the nuclear strategies of the Cold War look strange, even bizarre in their courting of global destruction. However, when seen from the perspective of the day, including the strongly held political ideologies of the principal participants, a certain logic appears, frightening perhaps, but nonetheless understandable. In the 1950s and 1960s, when the pace of nuclear development was most rapid, the United States saw itself in a life-or-death struggle with the evils of communism, a system that had already enslaved Russia, Eastern Europe, and large parts of Asia. Referring to the ultimate victory of world communism, Nikita Khrushchev shouted, "We will bury you!" in a speech at the United Nations. U.S. presidents Eisenhower and Kennedy were less dramatic but just as determined to contain the spread of communism and to make the benefits of democracy and free-market economics available to every nation. With such strongly held and diametrically opposite beliefs, there was an escalation of both rhetoric and weaponry, each side constantly attempting to gain the advantage while claiming that they were only seeking to maintain a balance of power against a threatening adversary.

In 1989, in the midst of a continuing arms race, the Soviet Union collapsed. While informed observers of communism saw its collapse as an inevitable result of inefficiencies and waste, few people expected the end to come when it did, or to occur so suddenly and so peacefully. Just months before the disintegration of the communist empire, military planners in the United States and other Western nations were busy studying Soviet military equipment and tactics and planning how we would defeat an invasion of Western Europe. The situation changed almost overnight in a way that hardly seemed real even as we watched history unfold on our television screens. The Berlin Wall was torn down under the eyes of guards who weeks before would have shot anyone who approached it. The Ukraine, the Baltic States, and the countries of Central Asia quickly seized the opportunity to assert their independence. Trainloads of Russian troops hurried eastward as the headquarters of once-feared secret police were ransacked by angry crowds.

One of the few things that *didn't* change during this turbulent time was that thousands of nuclear weapons continued to sit on alert status, aimed at targets that didn't seem to make sense in an era of peace dividends and glasnost. Deep within the bunkers where military strategy was plotted, nuclear attack plans were doggedly maintained. Submarines continued to sail with cargoes more destructive than all the weapons used in all the wars of history. Exercises were conducted to prove that bombers could be armed on short notice, ready to fly anywhere in the world. Simply put, no one knew how to change the system, or to what it should be changed.

THE END OF the Cold War was unusual in that it was less of a victory by the West than it was a self-generated collapse by the East. The Soviet Union was not "defeated" in any sense. It was never occupied by hostile forces, and no document of surrender was signed. While newly independent Ukraine, Belarus, and Kazakhstan willingly returned Soviet nuclear weapons to Russia, there was no question that the Russian Federation would continue to remain a potent nuclear power, if only because the deterioration of its army, navy, and air force meant that nuclear weapons were the only viable means for the defense of the country. There was discussion, to be sure, of why either side needed so *many* weapons when the adversarial relationship in which they were built no longer existed, but moves to reduce their number or to change the doctrine under which they would be used were slow in coming.

It was not until four years after the fall of the Berlin Wall that the United States took stock of its nuclear strategy and began to make changes. A study known as the Nuclear Posture Review (NPR) of 1993 took a comprehensive look at changes in the post–Cold War security environment and the role that nuclear weapons played in that environment. It recognized that, lacking the existential threat to America posed by the Soviet Union, fewer nuclear weapons were required to assure our future defense. However, uncertainty about the future suggested that the United States maintain a hedge—a reserve of nuclear weapons—should international events rapidly turn sour. Finally, to allay any suspicion that the United States might be abandoning a nuclear option, we reassured our allies that they should continue to rely on the security provided by our nuclear forces rather than develop their own.

The recommendations of the NPR seemed reasonable to many in the nuclear weapons community, but few changes actually took place. Perhaps the problem was that the momentum of the Cold War was simply too great, that there had been insufficient time to absorb the impact of the dissolution of the Soviet Union and the growing dominance of American conventional military power. The slow pace of change in nuclear policy drove the Congress to direct the Department of Defense to conduct a second Nuclear Posture Review in 2001, a study with the broad mandate of taking a fresh look at our nuclear forces in the light of a fundamentally changed world. Far from a recitation of the status quo, the NPR engaged some of the most creative thinkers in the defense community and came up with recommendations that would have been startling just a few years before. It abandoned the historical "strategic triad" of bombers, land-based missiles, and submarines to advocate a new one based on nuclear forces, defenses (e.g., missile defenses), and a responsive infrastructure or industrial base that could rapidly adapt to changes in the world. Whereas "strategic" used to be almost synonymous with "nuclear," the NPR suggested that nonnuclear conventional weapons could perform some of the missions currently assigned to nuclear bombs and missiles. This would enable a global response to impending danger without crossing the nuclear threshold and with vastly lower damage than would be inflicted by nuclear weapons. The

NPR also recommended a break from the policy of "mutually assured destruction" as a deterrent against aggression and advocated using modern technology to create a system of missile defense against small-scale attacks, again upping the threshold for a devastating nuclear response. Finally, noting that the pace of technology change was accelerating, it suggested that special attention be given to military technological preeminence—the United States could not afford to be surprised by a new type of weapon that would render us defenseless.

The NPR found that the United States needed far fewer nuclear warheads than when it faced strategic confrontation with the Soviet Union. However, it also recognized that the abrupt end of the Cold War left us with the wrong *types* of nuclear weapons for future missions—the highly sophisticated systems that were intended to destroy Soviet missile silos and bomber bases were ill-suited to respond to a geopolitical environment of proliferation and terrorism. Planners worried that the growing mismatch between future military requirements and existing weapons would lead to "self-deterrence," wherein the United States would be seen as *never* willing to use nuclear weapons, effectively destroying the very concept of deterrence.

With the NPR's recommendations for fewer nuclear weapons and a greater reliance on nonnuclear technologies, Admiral James Ellis, commander of United States Strategic Command from 2002 to 2004, commented, "The anti-nuclear groups should be writing the Congress in support of the NPR." Unfortunately, what was in fact an innovative and thoughtful report was all but killed by a series of lackluster and confused presentations by senior Pentagon officials and a classification mentality that kept even obvious conclusions a secret. Only a few official briefings of its principal findings were ever given, and several of those were by individuals who do not appear to have been adequately informed about the report. Rather than being embraced across a wide political spectrum as an agent of change, the NPR came to be seen by some as a mysterious instrument for perpetuating the Cold War. Even within the Pentagon the impetus for change was erratic and sometimes absent altogether.

The other failing of the Nuclear Posture Review—the extreme secrecy in which it was held—has yet to be corrected in a systematic manner. When I was the director of the Defense Threat Reduction Agency, the DoD organization tasked with activities related to weapons of mass destruction, I attended a meeting at the Pentagon on future nuclear strategy. In attendance were senior officials from all the relevant government organizations having a role in nuclear weapons. The briefer had hardly gotten started on his presentation when a heated argument broke out among the dozen or so people in the room. The contentious point was who was likely to be a future adversary of the United States, essentially at whom we would point missiles in times of war. After about thirty minutes I raised my hand and asked if it struck anyone else as odd that we, the leaders of the American nuclear weapons program, did not agree with or even know the policy that governed the use of the weapons in our care. If *we* were not authorized to know, then the hyper-secret policy might as well not exist.

A significant portion of the U.S. population, including many members of Congress, came of age after the end of the Cold War. To them, nuclear weapons are anachronisms, holdovers of a conflict read about in history books or seen in movies, some of which propagate false myths and rumors for lack of better information. Knowledgeable organizations such as the Center for International Security Studies have made numerous attempts to engage the Congress and the public in a debate about the future of nuclear weapons, but the press of other urgent business, not least of which has been the necessity of responding to the traumatic terrorist attacks of September 2001, meant that nuclear-related discussions were often poorly attended. Today the worst fears of the NPR's critics are coming to fruition: The United States is spending billions of dollars every year to maintain and refurbish weapons whose practical use military planners are hard-pressed to justify. Out of concern that *any* changes in the weapons in our nuclear arsenal (even those that would *reduce* the destructive force of our weapons) would result in a new arms race, the United States continues to maintain an arsenal vastly more powerful than we need. And at a time when the threat of terrorist attacks against nuclear facilities is a paramount defense concern, we have failed to implement modern security technologies that would greatly reduce the ability of any unauthorized person to use the components of a stolen nuclear device. It would be irresponsible if such policies continued after an informed and comprehensive debate. What is unconscionable is that such a debate has yet to occur nearly two decades after the end of the Cold War.

1

A Short History of Nuclear Weapons

The first decades of the twentieth century were a period of stunning advances in physics and chemistry. It seemed that every experiment, technical paper, and scientific conference revealed some new aspect of nature. Ordinary materials were shown to be constructed of complex combinations of ninety-two different types of atoms, which, with diameters of a few billionths of an inch, were considered the fundamental building blocks of nature. Later experiments showed that atoms were themselves constructed of a tiny central nucleus—with a diameter a few millionths of that of an atom—surrounded by a cloud of electrons, somewhat analogous to miniature solar systems.

By the 1930s, new tools were probing the interior of the atomic nucleus. Since the nucleus was far too small to be seen under the most powerful microscope, these methods often involved shooting beams of subatomic particles—neutrons, protons, and electrons—at materials and watching what happened. It was rather like shooting a gun at an automobile hidden in a tent—some of the bullets bounced back, some were absorbed, and some passed right through. By carefully analyzing the data, scientists could map out some of the basic properties of the car—or nucleus—without ever actually seeing it.

In 1938, German chemists Otto Hahn and Fritz Strassman reported on an astonishing phenomenon that they had observed while bombarding uranium atoms with neutrons. In some collisions the neutrons caused the nucleus to fission, or break into pieces, emitting several additional neutrons and a lot of energy in the process. The fact that uranium was radioactive—that it emitted neutrons—was not surprising. Marie Curie had earlier shown that many heavy elements emit some form of radiation. What was exciting—and frightening—about Hahn and Strassman's discovery was the notion that such a breakup could be *stimulated* by hitting a uranium atom with a neutron. Since each fission of a uranium nucleus released neutrons on its own, scientists immediately realized that a "chain reaction" could occur, wherein the neutrons released from one breakup would stimulate others, which in turn would stimulate others, and so on, with every such event releasing a sizable quantity of energy. If you put enough uranium together, the result could be a bomb of incredible destructive power, dwarfing anything that had come before. All previous explosives had used ordinary chemical reactions to produce their energy, reactions whose energy was measured in units of one or two "electron volts." The fission of one uranium nucleus produced *hundreds of millions* of electron volts of energy—suggesting that an entirely new class of super-weapon might be possible, millions of times more destructive than dynamite or TNT. The largest conventional bomb contained a couple thousand *pounds* of explosive, but preliminary calculations suggested that even a small atomic bomb would produce thousands of *tons* of explosive equivalent.

Even the earliest studies indicated that this task would not be easy, since the required chain reaction seemed possible only in an extremely rare form of uranium. Other elements such as man-made plutonium might be used, but they existed only in microscopic quantities and almost nothing was known about their properties. Still, given the implications of the discovery, many scientists were uneasy and even afraid.

While the implications of Hahn and Strassman's experiments were obvious to many American physicists, they were unknown to politicians and the military. How could the urgency of the situation be communicated to the national leadership without alerting other countries, some of which might be future adversaries of the United States? To maximize the impact of their warning, senior scientists decided to ask Albert Einstein, who had fled Nazi Germany and was working at the Institute for Advanced Study in Princeton, to write to President Franklin Roosevelt, alerting him to the danger and asking him to seriously consider a program to develop a weapon based on the newly discovered fission process.

Einstein and other senior members of the American scientific community feared that German scientists might already be working on the idea. Nearly every one of them had been trained in German universities, then the best in the world in physics, and they knew the capability of those who stayed behind with the Nazis. If German scientists were indeed working on their own atomic bomb, there was a very real possibility that they could hand Adolf Hitler a weapon that would instantly render all the armies and navies of the world obsolete.

Roosevelt accepted the argument and, over time, diverted

huge sums of money and materials to the development of the first atomic bomb. Factories and laboratories were constructed with astonishing speed, and the world's greatest concentration of scientific genius was focused on the single task of creating a practical atomic weapon. The Manhattan Project, as it was called, stretched from coast to coast, with nuclear reactors in Washington state, secret mountaintop laboratories in New Mexico, and sprawling manufacturing plants in Tennessee. Just twenty-three months after its inception, success was demonstrated in a spectacular explosion at Alamogordo, New Mexico, on July 16, 1945.

THE THREAT OF a German atomic bomb disappeared with the defeat of the Nazis in May 1945, but war continued to rage across the Pacific. Japan had vowed to fight to the last person, and Allied military planners soberly calculated losses of more than a million people in the invasion of the Japanese home islands. Could the atomic bomb be used as the ultimate "shock and awe" weapon that would stun the emperor and his advisers into admitting defeat? Would the loss of life, admittedly great in such an attack, eliminate the need for even greater losses during the expected house-to-house fighting across Japan? What was the best use of the bomb, if there was any, in bringing the war to the quickest possible conclusion?

Some scientists argued for a demonstration explosion on a remote island or at sea, an event that would be announced in advance to the Japanese government. Surely when the generals, admirals, and politicians saw the huge fireball and ensuing mushroom cloud—when they felt the scorching heat of a single explosion many miles away—they would see the futility of further resistance. But other scientists and military planners discounted the value of such a demonstration and pointed to the fact that the United States had only two atomic weapons; a demonstration might waste one of them to no effect. Only an attack on Japanese soil would produce a sufficient shock to end the war.

In the end, President Harry Truman made the decision to use the two available bombs in combat operations. Targets were chosen that had not suffered from the devastating bombing raids that had reduced Tokyo and other cities to little more than smoldering ruins. The hills that surrounded Hiroshima and Nagasaki would focus the effect of the blast, further increasing the destruction caused by the bombs.

ON THE MORNING of August 6, 1945, a B29 bomber nicknamed the *Enola Gay* approached Hiroshima at thirty-one thousand feet. There was no Japanese response to what appeared to be an isolated reconnaissance flight—the all-clear siren had already sounded after the uneventful departure of another plane, one that was checking the weather in advance of the atomic attack. Even if fighter planes were available, Japanese authorities figured that it was not worth the cost of employing them against such a small attacking force while Tokyo was being hit by thousands of bombs almost every day. Never in military history was there a greater miscalculation.

At 8:15 A.M., a single bomb fell from the *Enola Gay*. It took forty-three seconds to drop to nineteen hundred feet above the surface, a height chosen to maximize the damage produced by the expanding nuclear fireball. The first notice of its detonation was an intense flash—"brighter than a thousand suns"—that blinded anyone looking in its general direction. In an instant, this pinpoint flash expanded into an enormous

Mushroom cloud rising over Hiroshima after the atomic attack.

ball of fire that seemed to get hotter and hotter with each passing second. So intense was the flash and heat that buildings all over the city, many of which were of light wood construction, burst into flames. Within seconds a shock wave of staggering power crushed everything in its path. Thousands died before they realized what had happened and tens of thousands more were horribly burned and exposed to such intense radiation that their fate was sealed. The blast spared no one—men, women, and children, soldiers—even nine American prisoners of war were killed in the attack. Communication circuits were destroyed, delaying a report of the damage to Tokyo. When accurate information did get through to the capital, it was met with general disbelief. No one had ever seen an explosion of this magnitude, and there was no frame of reference in which it could be placed for understanding and analysis. By the end of the day, sixty-six thousand people would be dead and another sixty-nine thousand wounded.

Three days later, the United States demonstrated to Japan and the world that Hiroshima was not a one-off event by dropping its second bomb on the port city of Nagasaki. The destruction paralleled that of the first attack, with massive loss of life and total destruction of the central part of the city. While this

Remains of concrete and steel structures 0.85 mile from ground zero at Nagasaki.

was the last bomb the United States had ready for delivery, Japan did not know it. With the threat of losing several major cities per week, the "unthinkable" became thinkable, and the emperor and the fearsome Imperial General Staff accepted unconditional surrender.

CONTROVERSY HAS BEEN the constant companion of nuclear weapons. While the previous story is the standard American explanation of the end of the war, the Soviet Union painted quite a different picture. According to Moscow, the atomic bombings played a secondary role in convincing the Japanese to give up; the real reason was the declaration of war by the Soviet Union. The Japanese, communist historians wrote, recognized the futility of fighting against the massive Red Army and simply crumbled at the prospect.

Yet another school of thought insists that the Japanese were bluffing in their doctrine of defense at all costs and that they had already indicated a willingness to consider a cease-fire. If only given time, this theory went, the war could have been ended without the need for atomic attacks *or* ground invasions. As with all "what if's" in history, the use of atomic weapons against Japan will remain a subject of debate—the relevant fact for our discussion is that, after weighing all the evidence and listening to all the arguments, President Truman decided that the least of all dangers was to drop the bombs.

Many people think that the atomic bomb burst upon the world as a fait accompli, shocking friends and foes alike. In fact, its existence had been shared with Winston Churchill, and British scientists had been an integral part of the Manhattan Project from the start. At the Potsdam Conference in July 1945, Truman decided to bring Stalin in on the secret, a sign of goodwill that did not extend to any details of the new weapon. Expecting Stalin to reject the notion as preposterous or to get angry at not being told sooner, Truman was amazed when the dictator simply sat and listened with an enigmatic smile. Surely the Soviet leader simply didn't understand what was being said! In fact, Stalin knew more about the bomb than did the American president, since Truman had learned of the Manhattan Project only after the death of President Roosevelt, whereas Stalin had personally handled every piece of paper coming from the Soviet spies at Los Alamos. Stalin smiled because he had known more about the atomic bomb and for longer than did Truman!

IT TOOK INCREDIBLE effort to turn a simple scientific concept into a practical weapon. Hahn and Strassman's experiment suggested that if a sufficient quantity of fissionable material was assembled, a chain reaction would automatically follow. The problem was to obtain a sufficient quantity of the right kind of uranium or plutonium—both very rare materials—and to keep it together long enough for it to release its energy.

Uranium is a naturally occurring element found in small quantities around the world. But 99.3 percent of the metal in the mined ore consists of the isotope "238" (a designation of atomic weight), rather than the required "235" isotope. To make a weapon it is necessary to process many tons of metal-bearing ore to separate a few pounds of weapons-grade metal. Since the two types of uranium are chemically identical, the normal processes of chemical engineering were useless, and more esoteric methods had to be employed. Super-fast centrifuges were developed to separate the slightly lighter isotope, but their efficiency was so low, given the similarity in atomic weights, that thousands of them were required to get an appreciable output of fissionable metal. A diffusion process was developed that involved sending streams of uranium gas through thin membranes in the hope that the required separation would occur. Factories covering many acres were constructed in record time to implement these and other approaches.

The other element of interest, plutonium, had its own challenges. Plutonium does not occur in nature—it has to be made atom by atom in a nuclear reactor. Afterward it has to be separated from its host material through a set of chemical reactions that produce large quantities of highly radioactive nuclear waste. Since plutonium existed only in microscopic quantities at the start of the Manhattan Project, little was known about its material properties—how it could be machined, for example, and how readily a chain reaction would proceed in different configurations. All that could be done was to make intelligent guesses and hope for the best.

The next challenge faced by the scientists and technicians was how to keep the uranium or plutonium together long enough for it to produce a large amount of energy. Merely stacking blocks together would produce a chain reaction, but the resulting heat would melt the assembly, producing a flat puddle out of which neutrons would escape without contributing to the reaction. No steel or other material was strong enough to contain the immense energy output produced by even the earliest part of a nuclear explosion, so the idea of containing the metal in some type of pressure vessel was quickly abandoned. The solution to this seemingly intractable problem was found in the simplest of physical principles: inertia. The scientists realized that if they pressed a mass of material together very quickly, then inertia would prevent it from *disassembling* until a substantial amount of energy had been produced. A good analogy is what happens when a rubber ball is thrown against a wall—it takes an instant of time for the rubber to absorb the force of the impact and spring back in the opposite direction. If one could generate energy during the "bounce," the problem of containment would be solved.

Two techniques were used for the compression. In the first, two slugs of uranium were placed into opposite ends of an artillery barrel. Each slug was too small to produce a chain reaction on its own, but when they were fired together they formed a "critical mass" that could generate huge amounts of energy. It was such a "gun-assembled" nuclear device that was detonated over the city of Hiroshima.

Unfortunately, the gun method wouldn't work with plutonium. This man-made material emitted so much stray radiation that a chain reaction might start before the two slugs reached maximum compression. The result would have been a fizzle rather than a powerful explosion. To get around this problem, the implosion scheme was developed and tested in the first atomic explosion at Alamogordo, New Mexico. A ball

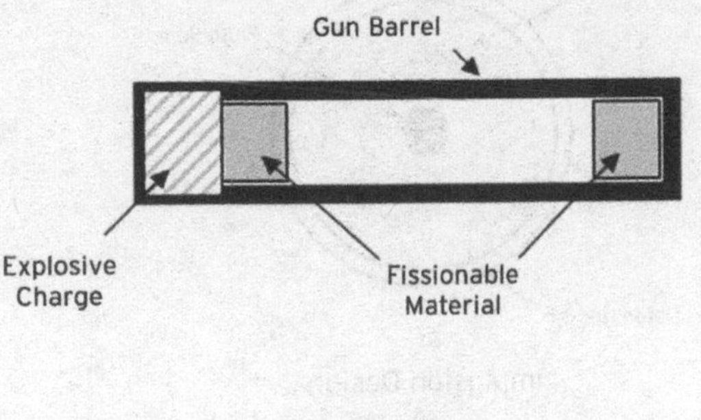

Gun Design

The gun-assembled scheme for a nuclear weapon in which one slug of uranium is fired at another one in an artillery barrel. (This design was used at Hiroshima.)

of plutonium was embedded at the center of a large sphere of conventional high explosive. Sophisticated methods were used to ensure that the high explosive detonated simultaneously at many points on its outer surface, so that a detonation pressure wave was created moving *inward* toward the plutonium. With many hundreds of pounds of high explosive pushing, the plutonium had nowhere to go and a chain reaction could be sustained long enough for large amounts of energy to be released. The second, implosion bomb was dropped on Nagasaki.

The atomic bombs dropped on Japan were primitive by modern standards: Each weighed more than four tons and produced yields (explosive output) between fifteen and twenty kilotons, that is, an amount of energy equal to the detonation of fifteen thousand to twenty thousand tons of TNT. (Modern nuclear weapons weigh hundreds of pounds and produce hundreds of kilotons of yield.) American scientists had many ideas for improvements that would have made weapons smaller

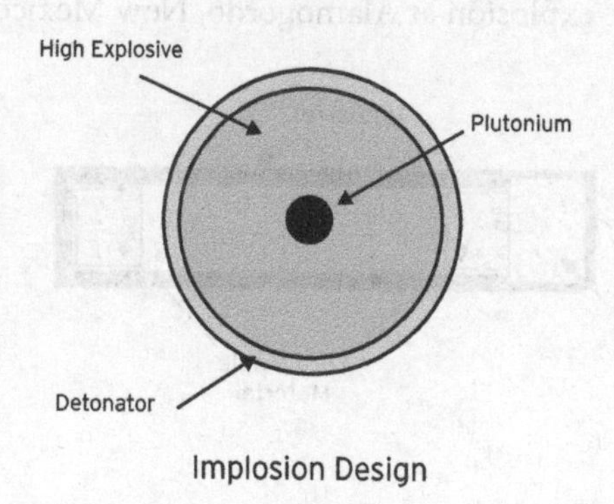

The implosion scheme for a nuclear weapon in which high explosives are used to compress a sphere of plutonium. (This design was used at Nagasaki.)

and more powerful, enabling several to be carried on airplanes or even on the newly developed ballistic missile. Hungarian émigré Edward Teller argued vociferously to move to the next level of destructive force by developing the hydrogen bomb, a device with hundreds of times the power of the atomic bombs dropped on Japan.

Another school of thought, exemplified by Robert Oppenheimer (the leading scientist on the Manhattan Project), argued for a pause in further development so that the world could come to grips with the staggering power that had been unleashed. He argued that atomic weapons might be placed under international controls to avoid the threat of any one country attacking another. After a contentious set of discussions, some of which were conducted publicly, Oppenheimer's security clearance was revoked. Teller had his way. America

Little Boy, the code name for the gun-assembled weapon dropped on Hiroshima at the end of the Second World War. The bomb was twenty-eight inches in diameter and ten feet long and weighed nine thousand pounds. It had the destructive energy of approximately fifteen thousand tons of high explosive.

Fat Man, the implosion bomb dropped on Nagasaki at the end of the Second World War. The bomb was sixty inches in diameter and ten feet, eight inches long. It weighed ten thousand pounds and had the destructive energy of twenty-one thousand tons of high explosive.

began a second nuclear crash program to develop a new class of weapons, if for no other reason than to keep pace with what was feared to be happening in Russia.

The hydrogen bomb is considerably more complex than the atomic bombs developed during the Second World War. Rather than using nuclear *fission*—the breaking apart of heavy atomic nuclei—it used nuclear *fusion*, the sticking together of light nuclei. (The term "hydrogen" bomb is actually a misnomer, since the fuel used in the H-bomb is actually a mixture of deuterium and tritium—isotopes of hydrogen containing a single proton and one or two neutrons—and other materials.) The challenge in the hydrogen bomb was analogous to the one faced by the designers of the first atomic bombs—how to get enough material together for long enough for fusion reactions to occur. Nuclei repel one another—they have a positive elec-

Robert Oppenheimer, who led the design of the first atomic bomb.

trical charge, and like charges repel one another—so to make two nuclei fuse together one has to apply pressures of millions of atmospheres and heat the material to millions of degrees. Only an atomic bomb had ever achieved these extraordinary conditions, so it was decided to use one of those as the first or "primary" stage to drive the "secondary" or fusion stage of a "thermonuclear" (i.e., very hot, involving atomic nuclei) hydrogen bomb. A test was hurriedly set up on a remote Pacific island, and the first fusion outside of stars was demonstrated. Rapid progress followed this experiment, with scores of tests conducted in the Pacific and later at the newly constituted Nevada Test Site, north of Las Vegas. Scientists worked around the clock to test different designs, improve their performance, and measure fundamental processes at conditions that could be produced only in a nuclear or thermonuclear explosion. This

Edward Teller, who argued for the development of the hydrogen bomb.

was a race, with many scare stories of Soviet accomplishments (some of which, while based on the best information available at the time, turned out to false) quickening the Americans' pace.

Scientists in all the nuclear weapons countries continued to explore new weapons concepts right up until the end of nuclear testing in 1992. We knew that the upper limit for a thermonuclear explosion was essentially infinite, but how small could a weapon be? A nuclear hand grenade was a standing joke, but a nuclear round for a short-range battlefield rocket was actually deployed. A potent atomic round was fitted into a six-inch artillery shell, a marvel of miniaturization, and a relentless struggle against weight was conducted to squeeze more destructive punch into ever smaller packages. (The driving force behind the push for smaller weapons was economic—smaller warheads saved the country thousands of dollars, but the smaller missiles that carried them cost many millions of dollars less than large ones.) But in reducing size and weight the weapons designers also reduced confidence—the more one optimized a system, the closer it came to the performance "cliff," beyond which it would fail to detonate. This is the same thing that happens in other complex machines such as automobiles. Racing cars can go much faster than the family station wagon, but they require much more attention and they don't always start on the first try. In every design there is a tradeoff.

The B83 strategic bomb carried on the B52 and B2 bombers.

Numerous advanced concepts for nuclear weapons were discussed and many were tested. Just as the atomic bomb was used to power the hydrogen bomb, the hydrogen bomb was used to power so-called third-generation concepts such as x-ray lasers and other directed energy weapons. President Reagan was infatuated with the notion of using nuclear explosives to power defensive weapons, and during the 1980s many programs were started to explore such concepts. However, after intensive work and the expenditure of billions of dollars, scientists realized that, even if they worked, such weapons would have little military utility, and after a time the third-generation programs were terminated.

The development of nuclear weapons in the United States effectively stopped in 1992 when President George H. W. Bush instituted a moratorium on further nuclear tests. The U.S. Senate refused to ratify the Comprehensive Test Ban Treaty that would have permanently halted nuclear testing, but the United States has nevertheless adhered to the moratorium and has used other methods to maintain its nuclear arsenal.

THE END OF the Cold War did not put an end to fears of nuclear proliferation. Western nations worried about what would happen to Russian nuclear expertise following the breakup of the Soviet Union. More than one hundred thousand people were involved in the Soviet program, many of them living in secret cities that appeared on no map and that were totally dependent on cash-strapped Moscow for their survival. Would Russian scientists be tempted to sell their knowledge to the highest bidder? Would this result in the proliferation of nuclear weapons to many countries, some of them hostile to the United States? Sigfried Hecker, then director of the Los Alamos National Laboratory, suggested in 1991 that we simply go and ask them—an astonishing suggestion for the time given the tight secrecy surrounding anything related to nuclear weapons.

Even more astonishing was the fact that Hecker's suggestion was accepted by the U.S. and Russian governments and resulted in a series of unclassified scientific collaborations between the American and Russian nuclear weapons laboratories. U.S. and Soviet scientists had known one another for some time, meeting at international conferences under the cover of performing basic research in other areas. The Russians gave false addresses and it had never been possible to visit their laboratories, but we knew that we had common interests in many areas of physics, such as the science of materials, lasers, high magnetic fields, and more.

I was involved in setting up many of the early collaborations, visiting the Russian secret cities dozens of times over a period of ten years. Prior to my first trip to Moscow, I bought a travel guidebook, complete with a foldout map of the city. What struck me immediately was that there were no target circles on the map—most of my experience looking at maps of the Soviet Union involved thinking of it as the enemy. This impression faded with time as I recognized that being at the Russian nuclear weapons laboratories was somewhat like looking into a mirror. Russian scientists had the same dedication to their country that I had to mine, the same attention to security, and the same concern about the potential proliferation of nuclear weapons to other countries.

Through detailed discussions with my counterparts, we developed a method called "step-by-step," wherein we would start with very modest objectives like sharing already published papers, progress through the discussion of what experiments we might do jointly, and finally conduct those experiments at one another's laboratories. The first joint project involved testing a super-high current electrical generator, one powered by high explosives, that the Russians had described in publications years before. Such generators are used to produce very high magnetic fields and to power other types of scientific experiments. Some Americans thought that the Russians were bluffing, that the device sketched in their publications could not work, so the first joint experiment consisted of repeating their experiment using modern American instrumentation to measure the output. It confirmed everything that the Russian scientists had claimed. Later, they told us that the figure in their paper was accurate so far as it went, but that it was not complete. They said with a subtle smile that there were several clever features that were deleted by Moscow censors. Some of their experiments looked like a Boy Scout troop assembled them—a mass of rough-cut metal, heavy wire, and glue. But they worked. What the Russians lacked in their ability to fabricate delicate components was more than compensated for by their ability to think of ways around the need for precision.

The opportunity to visit Russian nuclear weapons laboratories at the end of the Cold War gave Americans a new perspective on the Soviet side of the strategic equation. The following account of the Soviet nuclear weapons program was taken from near-verbatim notes of conversations I had with Yuli Khariton, the chief designer of the first Soviet atomic bomb, and others whom I encountered during my many trips to the Russian secret cities. Shut up for so long in their closed cities, unable to tell even their closest relatives where they were, the Russians were hungry to share their accomplishments, describe the challenges that they had overcome, and discuss the future with people who understood what nuclear weapons were really about.

AT THE END of the Second World War, the Soviet government saw itself surrounded by an American-led alliance dedicated to its containment—or even its destruction. Faced with a weapon of unprecedented power, Moscow saw little hope for survival if it could not match the West bomb for bomb, thus restoring a balance of power won at the expense of millions of lives during the war. The Soviet Union had to have its own atomic bomb, and have it as quickly as possible.

During the Second World War, spies within the Manhattan Project sent a steady stream of information on nuclear technology to Moscow. However, the recipients of these reports were never sure about its accuracy and were paranoid about the Americans discovering and plugging the leaks that were providing them with so much valuable data. To protect the spies' identities, only four people in the entire Soviet Union had access to all the information: Stalin, Lavrenti Beria (head of the KGB), Igor Kurchatov (a senior scientist), and one other (unknown to me) person. Curiously, Yuli Khariton, the chief designer of the first Russian atomic bomb, was *not* on the list, because of the fear that the information obtained by espionage was tainted to purposefully mislead the Soviets. Stalin thought it best to repeat everything from scratch and use the data obtained via espionage as the "answer in the back of the book" to check the results. This decision led to a type of hero cult surrounding Kurchatov in which senior scientists brought hard-won results to the bearded iconoclast only to have him reach into a safe behind his desk (where he kept U.S. design information) and say, "I'm sorry, comrade, but you're wrong," or "I agree with you—well done!" To the listener not in the know about the American data, it was little short of miraculous. "Kurchatov

The author standing next to a duplicate of the first indigenously designed Soviet atomic bomb, now on display in the nuclear weapons museum at the All-Russian Scientific Research Institute of Experimental Physics at Sarov, the Russian Los Alamos. This device had an explosive yield twice that of the first American bomb.

has already worked out everything on his own!" they thought. Not quite.

Soviet scientists knew how to do better than the heavy and bulky American designs used against Japan, but the first Russian atomic test was an almost perfect copy of the American implosion bomb dropped on Nagasaki. The only obvious differences were that it had a transparent nose containing a light for tracking its descent from the bomber and a set of radar antennas to measure altitude during that descent.

The decision to duplicate the American design rather than

use the more advanced Russian version was made by Stalin himself. He was intensely interested in the technology behind the atomic bomb, to the point of having the first plutonium core brought to his office in the Kremlin. "Can you divide it in half to make two smaller bombs?" he asked as he inspected the silver-colored sphere. "No, comrade, there is a precise lower limit to the amount of material needed for a working bomb." Stalin was averse to any unnecessary risk, considering it imperative to demonstrate that the Soviet Union was capable of matching anything in the West. He thought that the safest path was just to duplicate the American design. "How do I know that it will work?" was the last question he put to the scientists assembled in his office. "Put your hands around it, comrade, and feel its heat," was the reply. (The natural radioactivity of plutonium makes it warm to the touch.) Stalin did so and replied simply, "Shoot it."

Affairs took a more practical turn outside of Stalin's office. KGB head Lavrenti Beria informed the technical staff that rewards and punishments commensurate with responsibility would be dispensed immediately after the first test. If successful, the chief designers would be made Heroes of Socialist Labor. If unsuccessful, they would be shot on the spot by military firing squads positioned at the control bunker for just that purpose. Lesser scientists and engineers received lesser promises—a new car or ten years in a Siberian labor camp. There was a definite motivation to succeed.

Their first test was indeed a success, shocking the world and particularly the American intelligence community, which had predicted that the Soviets would require several more years to reproduce the technological feats of the Manhattan Project. The explosion had about the same yield—twenty kilotons—as the American design, not surprising given that it was an exact copy. The West would have been even more worried if it had detailed data on the second Soviet test. Having demonstrated a nuclear capability, the Soviet government permitted its scientists to try their own design. It produced forty kilotons of yield, twice that of the American design.

From the very start, the Soviets had a different approach to nuclear weapons development than the Americans. While the Americans tried many different experiments to see what was possible, the Russians were always focused on practical weapons that could quickly be sent to their troops. Andre Sakharov, one of the best of the Russian nuclear weapons scientists, had his own ideas for hydrogen bombs and quickly applied them. It was, in fact, the Russians and not the Americans who demonstrated the first practical weapon using the principles of nuclear fusion, one of several occasions during the early days of the Cold War when Soviet nuclear technology was ahead of its American counterpart. Part of the reason for this prowess was the method that the Russians used to recruit promising young scientists for their nuclear program. Sakharov and other senior scientists periodically visited Soviet universities to administer a test to scientists and engineers nearing graduation. (The students were never told who the mysterious visitors were; they were simply two older men dressed in tattered sweaters who seemed to know a lot about physics.) The very best candidates would be given a choice: Go to a secret location to work with the best scientists in the Soviet Union, using the very best equipment available, or be labeled an enemy of the people. Most chose the former option and remained in the secret cities for the remainder of their lives. In addition to the attraction of the technical work, they enjoyed unparalleled access to luxury goods from around the world thanks to their access to premium warehouses established for the Moscow elite. Whereas most American nuclear weapons designers migrated to other things, some going back to pure science and others to management positions, Russian scientists stayed on the job, steadily improving their skills despite lacking the advantages of supercomputers and other advanced equipment.

In the mid-1950s, Nikita Khrushchev decided to demonstrate Soviet mastery of nuclear technology by showing that he could create a weapon of almost unlimited destructive power. He chose one hundred megatons as the target yield, many times greater than what either the United States or the Soviet Union had yet achieved. Andre Sakharov argued against such a titanic explosion, saying that it could cause lasting global environmental damage from radioactive fallout. Khrushchev ultimately relented, and the design was diluted to "only" sixty megatons. Weighing in at more than thirty tons, the device was too big to fit in the biggest Soviet bomber and was instead carried on a special bracket underneath the aircraft. An enormous parachute was fitted to the bomb to allow the plane time to escape the fireball. As an added precaution, the wings and fuselage were coated with reflective paint to keep them from melting, and the pilot and copilot each had two feet of lead shielding behind their seats to limit their radiation dose. Even so, both were told that they were unlikely to survive the mission and that they would be made Heroes of Socialist Labor for their sacrifice. (They both died painful deaths within a year.)

Yuri Trutnev, who worked "side by side" with Sakharov on the design of the monster weapon and who likely had the greater influence on its design, described the test to me during a visit to his laboratory. While the detonation occurred over the island of Novaya Zemlya, the Soviet test site north of the Arctic Circle, the control bunker was well south on the Russian mainland, partly to protect the scientists should something go wrong. "When the plane approached the target," Trutnev told me, "we heard the conversation between the control center and the pilot until, at the final moment, everything went to static. The northern sky lit up with a terrific brilliance that went on and on. All radio communications were cut by the intense x-ray pulse ionizing the atmosphere for hundreds of kilometers around the blast." (Radio waves cannot penetrate ionized gases.) "We waited and waited, wondering what could possibly

Left to right: Yuri Trutnev, who designed the largest nuclear explosive ever detonated; Yuli Khariton, the chief designer of the first Soviet atomic bomb; the author. Photo taken in the nuclear weapons museum of the All-Russian Scientific Research Institute of Experimental Physics.

be taking so long for the signal to be restored—it was only later that we realized how *much* ionization had been caused by the explosion. In fact, the bomb was so powerful that it blew the top off the atmosphere."

The bomb detonated over Novaya Zemlya was the largest ever constructed by any nation. It was more than four feet in diameter, compared to the one to two feet typical of modern strategic weapons. A spare set of parts associated with its thermonuclear stage was inscribed as a world globe and now sits outside the House of Art at the second Russian nuclear laboratory near the Urals, something that the scientists there find weirdly humorous.

Yield was not the only criterion for advanced weapons. Both Russia and the United States pursued designs for "clean" nuclear explosives that produced less fallout than typical weapons. One of these was the so-called neutron bomb, which released a substantial part of its energy in the form of neutrons—the idea being that its effects were not as long-lasting as those of conventional nuclear explosions. Recognizing the potential impact of such weapons on their massive tank and infantry formations in Western Europe, the Soviets launched a brilliant publicity campaign in the 1980s against American deployment of the neutron bomb, claiming that this was the ultimate "capitalist bomb" that killed people but left property intact. In fact, the Russian laboratories were working on the same kind of weapon for the very same reasons as the Americans. During a visit to one of their nuclear weapons institutes I was shown a sign, written in English, that was used during marches protesting American nuclear weapons deployment in Europe. "We won that battle," observed my host.

While Soviet nuclear weapons science was on a par with America's, Soviet scientists were continuously challenged by problems in engineering and manufacturing. Some of their elegant concepts were simply too complex for their manufacturing plants to build, with the result that most Soviet weapons were relatively crude-looking to Western eyes. Soviet bombs were bigger, partly because their missile warheads were less accurate, and few of the scientists' advanced concepts made it into the stockpile. This conservatism was actually encouraged by the Soviet military, who distrusted the arrogance of the scientific staff to the point where they insisted on having their own personnel present during the assembly of each weapon.

MERELY HAVING A nuclear explosive does not constitute a nuclear deterrent—you also need the means to deliver it to the target and have it detonate at the intended time and place. The earliest nuclear weapons were massive devices—they weighed several tons each and required last-minute attention before detonation. The biggest bombers of the 1940s could at best carry a few of them, and such massive aircraft were vulnerable to anti-aircraft fire, fighter planes, and other dangers on the way to their target.

To circumvent these problems, military planners decided to use ballistic missiles as delivery vehicles on the theory that their sheer speed made them virtually impossible to intercept. While the American scientist Robert Goddard had invented many of the key concepts of long-range missiles, the U.S. government did not recognize the potential of his work. It was only when the Germans developed the V2 missile to attack London in the Second World War that the military began to take the new technology seriously. In the closing days of the war in Europe, both the United States and Russia rushed to collect as many German rocket scientists as they could, along with any relevant hardware and rockets. Both used the Germans' expertise as a springboard to their own programs, which were designed as a mix of unclassified space exploration and highly classified weapons development.

Early warheads on ballistic missiles often had very high yields, sometimes in the megaton class. The accuracy of early missiles was so poor that a huge explosion was required to ensure that their targets would be destroyed. When asked what constituted "accuracy" for such a missile, one admiral in charge of early submarine missile development replied, "It detonates over enemy territory." High yield wasn't a case of destruction for the sake of destruction, but a necessity given the limitations of current missiles. Missile accuracies have greatly improved since the last nuclear weapon was introduced into the U.S. stockpile in the 1980s, but there has been no political support to follow through with a reduction in the yields of our nuclear weapons.

The Soviets were obsessed with the need for secrecy, so much so that they sometimes refused to allow people working on the same program to talk to one another. Andre Sakharov remembered the day that a team of nameless strangers came to talk with him about hydrogen bombs. "How big would one of those be?" they asked. Sakharov thought for a moment, added a healthy dose of conservatism (errors could have fatal consequences in those days), and gave his answer. What he did not know was that the team questioning him was from the rocket design bureau. They took his numbers literally and set about to design enormous missiles to carry the massive warhead. If Sakharov had known why they were questioning him he might have given a more reasonable answer, saving the Soviet Union enormous amounts of money in its strategic missile program. In contrast, American weapons designers worked hand-in-glove with the aerospace industry so that much smaller and much less expensive missiles were required.

DURING THE COLD WAR, nuclear weapons were considered the ultimate weapons for attack and defense, and no expense was spared on their development and manufacture. Ballistic missiles offered the advantage of speed, but they were vulnerable to attack while they were on the ground. What if the Russians attacked and destroyed all our missiles? One answer was to bury them in hardened concrete and steel silos, but the better solution was to put them on submarines. In the vast expanses of the oceans, nuclear weapons could hide until needed.

Admiral Hyman Rickover had been championing the promise of nuclear-powered submarines, noting that with their nearly inexhaustible power sources, such vessels could stay submerged for months at a time. A major effort was expended by the U.S. Navy to miniaturize nuclear-capable missiles so as to enable a submarine to carry them, paralleled by an effort at the nuclear weapons laboratories to make a potent warhead small and light enough to be carried long distances on the diminutive missiles. The result of Rickover's vision was a fleet of nuclear-powered submarines that could prowl undetected across the seas, ready at a moment's notice to send a hail of warheads flying toward the enemy. While the Soviet Union developed its own nuclear submarines, this was a case where the technological edge of the United States paid off in safety, efficiency, and performance. Soviet submarines were quiet and hard to detect, but they suffered frequent radiation leaks and sometimes had to be towed to their duty station to limit the exposure of the crew to dangerously high levels of radioactivity.

THE UNITED STATES and Russia are the world's nuclear superpowers, each having roughly comparable knowledge of this unique class of weaponry. Each constructed several tens of thousands of nuclear weapons—more than enough to obliterate all civilization on the planet and, as one analyst observed, "to make the rubble bounce." However, other countries were intent on building their own nuclear arsenals, if not for defense then for the national prestige associated with being a member of the "nuclear club." Britain, a partner in the Manhattan Project, constructed a laboratory at Aldermaston and demonstrated its own weapons designs in testing done in the Pacific, Australia, and later at the American test site in Nevada. France struggled to achieve a nuclear detonation of any kind, but quickly learned the secrets of high-performance design and has produced weapons equal in sophistication to those of the nuclear superpowers.

France's difficulty in creating a first-generation nuclear capability suggests that we should doubt assertions that the information required to make a nuclear weapon is freely available. On a tour of the French nuclear weapons museum, I asked my host, "Why on earth did you do that?"—referring to certain features of their early designs. His answer was, "We just didn't know any better." We will discuss nuclear proliferation in a later chapter, but it is worth noting here that even a technologically advanced nation like France, with outstanding scientific and engineering talent, was hard-pressed to create a practical nuclear weapon.

China received substantial help from Russia in creating its nuclear deterrent, almost to the point of being provided a complete set of plans for a weapon. Only at the last moment did Moscow demur from sending the blueprints, a perceived betrayal that is still fresh in the minds of Chinese weapons personnel. As a constant reminder of the need for an independent program, the date of the Russians' pull-out is marked on Chinese security badges. China's weapons are relatively crude compared to those of the other four declared nuclear powers but are more than adequate to level a large city.

THE NUCLEAR NONPROLIFERATION TREATY (NPT) of 1970 was intended to limit the nuclear club to five members—the United States, Russia, Britain, France, and China—while providing the peaceful benefits of nuclear technology to any country that agreed not to develop weapons. The same uranium and plutonium used in nuclear weapons can be used in nuclear power stations, generating electricity without the carbon emissions associated with fossil fuels. Radioactive isotopes are used to diagnose and treat cancer and as power sources in satellites on multiyear explorations of deep space. Some glow-in-the-dark watch dials contain minute amounts of radioactive materials, as do household smoke detectors and devices designed to map oil and gas wells. Many countries want to take advantage of these technological opportunities and bristle at the suggestion that any interest in things nuclear necessarily means that they are developing nuclear weapons. The NPT has been generally successful, although in recent years India, Pakistan, and North Korea have developed nuclear weapons, and others, most notably Iran, seem on a path to do so.

2

How Did We Arrive at the Theory of Mutually Assured Destruction?

Few concepts in military strategy are more puzzling than the notion of mutually assured destruction. More than frightening, it sounds irrational or even insane to plan for the total obliteration of society, an apparent contradiction of the very motivation for war. Why would anyone propose a strategy in which *both* sides lose? For all its disturbing and apparently illogical aspects, the doctrine of mutually assured destruction, or MAD as it came to be called, involved some of the best strategic thinkers of the twentieth century, including presidents Dwight Eisenhower and John Kennedy, and Defense Secretary James Schlesinger. It reflected the uncertain times in which decisions were made, the novelty of the weapon, and, quite simply, an inability to come up with anything better.

The origins of MAD date to the strategy of massive aerial

bombardment developed in the Second World War. There was little doubt in the spring and summer of 1945 that the Allies would eventually prevail over the Japanese. The only question was how long it would take, how many lives would be lost, and how much destruction would be wrought before the Imperial General Staff finally capitulated.

The strategy to defeat Japan took two parallel paths. First, the Allies waged a relentless campaign to destroy the Japanese war machine, both its troops and the industrial base that supplied them. Simultaneous American thrusts led by Admiral Chester Nimitz (from the east) and General Douglas MacArthur (from the south) were eliminating what remained of the Japanese army, navy, and air force, and were cutting off the home islands from much-needed oil, rubber, and other vital resources. At the same time, under the direction of General Curtis LeMay and Colonel Tommy Powers, a withering strategic bombing campaign was conducted to eliminate war industries, transportation networks, ports, and anything else that might prolong the hostilities. But even the astonishing level of damage caused daily by hundreds of B29 bombers was considered too slow a process to bring to terms a government that had vowed to defend its homeland to the last drop of Japanese blood. So, in parallel with attacks on industrial and military sites, American generals decided to bomb the civilian populations of major cities to destroy their morale and hence hasten an end to the war. It was within this context of purposeful and massive destruction that the decision to use the first atomic bombs was made. Lacking any new policy to govern the use of atomic weapons, they were simply inserted into the existing policy of mass aerial bombardment.

So different was the atomic bomb from all weapons that preceded it that even seasoned military officers refused at first to believe that a single explosion could cause such destruction. This is illustrated in an anecdote related by Harold Agnew, a crewmember on one of the planes involved in the Hiroshima attack and later director of Los Alamos Scientific Laboratory. Immediately after the cessation of hostilities, Agnew was assigned to Tinian, the island from which the attacks were launched, to brief visiting generals and admirals on the bomb and its effects. He used the box that held the plutonium core of the Nagasaki bomb, about the only thing that was left after the bombs had been dropped, as a prop. During one of Agnew's talks the ranking general scowled and said, "Son, you may think that a city could be destroyed by what was in that box, but I don't have to believe it," after which he stood up and walked out.

To other officers, the fact that a single bomb could achieve a level of destruction that had previously required thousands of conventional bombs was only a quantitative distinction, a large but nevertheless understandable advance in military capability. More than one hundred thousand people had been killed in the firebombing of Tokyo, more than were killed in either of the atomic attacks, so the sheer level of destruction was not demonstrably different. Many military thinkers viewed atomic bombs as just another weapon in the arsenal, usable against any future enemy that might threaten the interests of the United States.

President Truman thought otherwise, as has every U.S. president since. He saw nuclear weapons as a *qualitative* shift in warfare, a transition point that separated the past—when wars were frequent but survivable—to a future when conflict might end civilization itself. Truman saw the use of nuclear weapons as a presidential decision and insisted that their development remain in civilian hands. Congress created the Atomic Energy Commission in 1946 to oversee their development, and in the autumn of 1948, Truman formalized presidential authority over atomic weapons in a national security policy memorandum.

IMMEDIATELY AFTER THE war there were discussions of international controls on nuclear weapons and the special forms of nuclear material—uranium and plutonium—that powered them. Some suggested putting them under the authority of the newly founded United Nations, to be part of an international peacekeeping force. Here Truman had the support of the Joint Chiefs of Staff, who saw the proliferation of nuclear weapons as the only threat to American military superiority.

Such utopian dreams were shattered as the Soviet Union rejected any form of international controls on atomic energy, believing that such policies would only cement the superior position of the Western powers. In 1948, Soviet occupation forces in Germany cut off access to the American, British, and French controlled sectors of Berlin, creating what Winston Churchill described as an "Iron Curtain" separating East and West. That same year, the United States created the Strategic Air Command (SAC) under the leadership of General Curtis E. LeMay, who had previously designed the campaign of massed air attacks on Japan. SAC was responsible for all types of strategic bombing, but it quickly focused on the unique aspects of nuclear warfare.

The American monopoly on nuclear weapons was broken a year later when, in 1949, the Soviets conducted their first nuclear test, well ahead of American intelligence estimates. It was no longer a question of *whether* the United States should have nuclear weapons, but *how* those weapons would affect an evolving geopolitical struggle between two irreconcilable political systems. Seeing the Soviet Union as the most likely adversary of the future, General Curtis LeMay and his staff at SAC developed nuclear war plans that included attacks on war-fighting industries, transportation networks, and associated national infrastructure—a direct continuation of the approach he used against Japan. More than two hundred critical targets were identified, including most of the major cities of the Soviet Union, and an intense manufacturing program was launched to supply the nuclear weapons required for such attacks.

At the same time, planners began to assess the effect of a Soviet strike on the United States by long-range bombers and commercial ships, the latter a harbinger of modern worries over terrorist weapons on ships. Tactics and weapons were developed to protect the United States from air and sea attack. The extraordinary destructive power of nuclear weapons meant that anything less than 100 percent success in their interception—something that was already considered impossible—would result in catastrophe.

Pressure was also building for the consideration of nuclear options in otherwise conventional (i.e., nonnuclear) wars. When North Korea marched south on June 25, 1950, the Pentagon feared that all might be lost on the peninsula unless overwhelming force could be brought to bear, and quickly. With hardly enough troops to put up a delaying action, American generals argued for the use of atomic weapons as practical weapons of war, an equalizer against massed Chinese attacks. But a closer analysis revealed that there were few targets in Korea that could not be destroyed with conventional weapons, and wiser minds noted that the use of nuclear weapons against supporting Soviet or Chinese bases could trigger a global war that the United States was ill-prepared to fight. This was the first example of what would become a persistent dilemma in nuclear strategy: The risk of using nuclear weapons could easily outweigh their military benefit. They were not practical instruments of war like battleships and tanks. Their use could trigger an escalation of conflict that was far more destructive than the conflict they were designed to stop.

Nevertheless, General LeMay and other military commanders continued to argue for a first strike against the Soviet Union, a "preventive war" that would settle the nuclear standoff once and for all. Truman was opposed to a first-strike option, but he did authorize contingency planning for war with the Soviet Union, including a fundamental change in how nuclear weapons were to be employed in a strategic conflict. Previously, the highest priority targets were military units and war industries. Now, a "counterforce" strategy was developed that put priority on the destruction of enemy nuclear weapons that could strike the United States. The goal was not to destroy the Soviet Union, but to prevent the Soviet Union from destroying the United States.

After the Russians demonstrated that they too had the atomic bomb, Truman realized that he could no longer unilaterally control nuclear technology. He authorized research into the full spectrum of atomic munitions, from those with yields of only a few tons intended for battlefield applications to megaton behemoths for strategic "city busting." Edward Teller and conservative elements in the scientific community pushed tirelessly for the United States to start work on the hydrogen bomb. They argued that the Soviets were probably already working on their own "super" bomb and that the United States could not allow itself to be found flat-footed in an arms race. (Russian nuclear weapons designers gave the identical reason for their hydrogen bomb program—the Americans were already well along and the Soviet Union could not place itself at a strategic disadvantage.) The first American test of the hydrogen bomb, code named "Mike," was conducted in November 1952, graphically demonstrating the almost limitless destructive potential of the H-bomb. It seemed practically impossible to halt the development of new types of atomic weaponry.

DWIGHT EISENHOWER INITIATED his own review of nuclear weapons policy when he was sworn in as U.S. president in 1953. He understood from personal experience the tenuous position of the Western powers in Europe and saw nuclear weapons as the essential balancing factor to massive Soviet troop concentrations deployed on the German border. The United States could not afford to match the Soviets soldier for soldier and tank for tank; nuclear weapons provided a counter at much lower cost. As the Iron Curtain settled into place, American planners began to talk of containment of Soviet expansionism and perhaps an eventual rollback in which the occupied countries of Eastern Europe would be freed from communist domination. The notion of "deterrence" was refined to emphasize the role of nuclear weapons in preventing any provocative action on the part of the Soviets—any move that threatened the United States or its allies would bring a swift and devastating response.

Eisenhower was intensely engaged in all aspects of nuclear strategy, from what policy should govern the use of weapons to how nuclear science might be used for peaceful purposes. Reviving interest in international controls, he announced an "Atoms for Peace" program in his first year in office, which aimed to begin an international dialogue on the future of atomic energy. The stated goal of the program was to find ways to make the peaceful uses of the atom available to other countries while discouraging them from pursuing weapons. Bringing more countries into the nuclear fold was a calculated risk, all the more remarkable in that it occurred at the height of the Cold War.

WHILE EISENHOWER WAS holding out an olive branch, his secretary of state was brandishing a stick. On January 12, 1954, John Foster Dulles gave a speech to the Council on Foreign Relations in New York that outlined a new policy of massive retaliation to Soviet aggression. In an apparent abandonment of the idea of nuclear weapons as unique instruments of destruction, a National Security Council directive issued the previous year stated that "the United States will consider nuclear weapons to be available for use as other munitions," effectively creating a policy of ambiguity that was designed to deter any potential adversary from attacking.

Dulles's speech was actually penned by Eisenhower, who wanted to present a new approach to the world (Atoms for Peace) while maintaining a strong line against communism. He believed that the Soviets did not want nuclear war any more than did the United States. He was especially concerned that such a war, once started, might be difficult or impossible to control. The first nuclear weapon used would inevitably lead to a second, a third, and so on until there was a massive exchange of hydrogen bombs on each side. To prevent disastrous decisions from being made in the heat of an international crisis, he directed that a Single Integrated Operational Plan (SIOP) be developed to coordinate all nuclear war fighting by American forces. Previously, each military service had its own nuclear plans, some of which interfered with one another. Eisenhower forced the services to accept central planning and control of nuclear weapons, always under direct presidential authority, as a means of ensuring that they followed his strategic policy.

By 1955, the essential questions that would govern all future nuclear debate were firmly established. What is the role of nuclear weapons in fighting limited wars, and how can escalation to full-scale nuclear war be avoided? What is the role of defenses—anti-aircraft missiles and ballistic missile defenses—in nuclear war planning? Finally, how can deterrence be maintained without *provoking* a nuclear war by stimulating an enemy first strike? The fundamental tension affecting nuclear policy has always been between a commitment to use them if absolutely necessary and a hope that their destructive potential will *never* be unleashed. To say that you would *never* use a weapon renders it ineffective as a deterrent to aggression—the adversary knows in advance that you will not shoot and acts accordingly. But to *plan* on using a nuclear weapon in anything other than the most extreme circumstances might trigger a war in which both sides would lose. Such contradictions have been a constant problem for nuclear planners.

THE HYDROGEN BOMB and the development of intercontinental ballistic missiles (ICBMs) were parts of an accelerating arms race in which each side considered its very survival dependent on keeping up with, or being just ahead of, its adversary. East-West distrust was fueled by lack of credible information on what the other side was doing, a deficiency that contributed to warnings of a "bomber gap" and later a "missile gap" between the United States and the Soviet Union—gaps that could spell strategic defeat if left uncorrected. The launch of Sputnik in October 1957 only exacerbated fears that the Soviets would soon be able to deliver nuclear weapons anywhere in the world. The tiny satellite itself posed no threat to the United States—it carried neither weapons nor spy cameras—but its very presence in the night sky demonstrated that the rules of strategic war had changed. If the Soviets could put a satellite in orbit, they could use the same missile to put a warhead on an American city, and do so in less than an hour.

A RAND Corporation study done in the early 1950s warned that American bomber bases were vulnerable to a Soviet first strike that could destroy most of our nuclear stockpile, rendering the United States incapable of mounting an effective counterstrike. Two solutions were implemented to deal with these threats. First, a new generation of surface-to-air missile batteries were placed around the country's borders to defend against incoming Soviet bombers. Second, a new emphasis was placed on the delivery of nuclear weapons by ballistic missiles, ensuring that American warheads could break through the formidable Soviet air defenses. This was the beginning of what came to be known as the "strategic triad" of bombers, land-based missiles in hardened silos, and sea-based ballistic missiles on submarines. Bombers could be controlled right up until the time they dropped their bombs, as opposed to missiles that, once launched, automatically flew to their targets. However, bombers could crash or be shot down. Land-based intercontinental missiles were not vulnerable to enemy air defenses and, in principle, could be controlled from Washington, but they were fixed in location and hence could be destroyed by a massive enemy first strike. Submarines offered the same advantages of land-based missiles and were virtually impossible to detect and destroy, meaning that some would survive to inflict a devastating counterstrike on the Soviet Union. But communication with submarines could be a problem in a time of crisis. Each leg of the triad had its advantages and disadvantages; together they represented an almost indestructible implementation of the theory of deterrence.

Since sea-based missiles were much less accurate than bomber-delivered weapons or ballistic missiles launched from fixed land locations, they necessitated a shift from the evolving "counterforce" strategy, one that focused on eliminating Soviet nuclear capabilities, to a "countervalue" strategy that targeted large population centers. The Polaris missile, itself a modern engineering marvel, was capable of delivering a compact but powerful nuclear warhead over thousands of miles, but it could not ensure that it landed with sufficient precision to destroy Soviet missiles or even bombers that might be located in hardened shelters. The only use for such weapons was against very large "soft" targets such as cities. A shift in policy thus came about as a result of the capabilities and limitations of technology.

Thinking that restraints on nuclear testing were a way to moderate the development of new weapons, Eisenhower reached an agreement with the Soviet Union in 1958 to stop all testing of nuclear weapons. He established the U.S. Disarmament Agency to develop and implement new arms control measures, a first step toward dismantling the massive arsenals that had already been built. It is remarkable that a former general took major steps to reduce what he saw as a growing nuclear threat, especially at a time of intense competition with, and distrust of, the Soviet Union.

EISENHOWER'S POLICIES FOR controlling the use of nuclear weapons in any future conflict still focused on a single massive attack against the Soviet Union. When President John Kennedy was briefed on the country's nuclear war plans after taking office in 1961, he was astonished at their rigidity and destructiveness. Surely there must be something better than an "all or nothing" strategy that launched the full arsenal in one desperate bid for victory. Kennedy was appalled by the incredible devastation that would ensue from a nuclear exchange, including the projected hundreds of millions of civilian casualties. With Kennedy's concurrence, Secretary of Defense Robert McNamara proposed a "no cities" strategy in which population centers would be avoided in favor of military targets, especially nuclear missiles that could threaten the United States. Recognizing that both sides had to adopt this new approach for it to have any value, he proposed discussions with the Soviets on the rules that might govern a future nuclear war. Unfortunately, the Soviet Union rejected the policy, reinforcing the arguments of General Curtis LeMay (who by that time was chief of staff of the air force) that the only way to deal with the complexities of nuclear war was to mount a massive preemptive strike to destroy the Soviet Union.

Kennedy rejected LeMay's gamble, and a new SIOP was developed to implement what came to be known as "flexible response," a limited use of nuclear weapons well short of the massive exchanges of previous plans. The United States would use nuclear weapons only to achieve urgent military objectives. The new SIOP included a reserve force, ending what was essentially a policy of launching everything in one massive strike against the enemy. Even after a "military exchange," there would be sufficient weapons left over to destroy Soviet society, hence maintaining an assured destruction element to deterrence.

The magnitude of destruction that was "assured" by this reserve force was breathtaking: McNamara projected that a third of the Soviet population and half of its industrial capability would be eliminated by an American counterstrike, a potent deterrent to any rash act on the part of the Kremlin. With these frightening statistics in mind, McNamara tried to stem the unchecked arms race by insisting that the United States needed only enough weapons to accomplish preestablished objectives—there was a point at which "enough was enough." More weapons just for the sake of more weapons could actually make the country *less* safe, since a massive nuclear stockpile could frighten a potential adversary into attempting a devastating first strike of its own. Also, the rising cost of nuclear arms meant that fewer resources were available for other vital national security programs.

THE SOVIET UNION broke the moratorium on nuclear testing in 1961 with a rapid series of explosions at their test site on Novaya Zemlya, an island north of the Arctic Circle, and at a desert test range in Kazakhstan. The United States quickly followed suit, and the arms race was back in earnest. However, negotiations aimed at limiting the environmental damage caused by nuclear explosions continued, and just before President Kennedy was assassinated in 1963, the Limited Test Ban Treaty entered into force, an agreement that outlawed nuclear testing in the atmosphere, in the ocean, or in space. Only underground tests were to be permitted, a compromise that permitted arms development with much less radioactive fallout. Beyond its immediate goals, the Limited Test Ban Treaty set a precedent for future arms control agreements between the United States and the Soviet Union.

By the early 1960s, the nuclear arsenals of the two superpowers had reached staggering levels, with many thousands of weapons on each side. B52 bombers flew circular patterns just outside Soviet airspace, ready to pounce upon preassigned targets on receipt of a properly coded message from the White House. More bombers sat at the ends of runways, engines running, in case a Soviet strike was detected. Missile crews stood alert twenty-four hours a day, and missile-carrying submarines remained submerged and undetected for months at a time. Nuclear warheads appeared on almost every conceivable weapons system, including naval torpedoes, field artillery, and even backpack-sized demolition munitions. Technology was developing so rapidly that a weapon remained in the stockpile only a few years before it was replaced by a newer and more efficient model. Higher yield was not the only objective, as safety and security were major concerns in new weapons. Designers worked to ensure that a bomb would go off only when it was intended to do so and would not explode if involved in an airplane crash, a fire, or other accident.

EUROPE WAS CONSIDERED a flash point in East-West relations. Since the end of the Second World War the Soviet Union and United States each stationed tens of thousands of troops in East and West Germany. Western Europe feared that the Soviet troop concentration was a prelude to invasion, a completion of the land grab that Moscow undertook at the end of the war. The North Atlantic Treaty Organization (NATO) was formed as a bulwark against possible Soviet aggression, an alliance that linked most European countries with the military might of the United States. (France withdrew, not wishing to place its defense in the hands of others, and developed its own nuclear deterrent to back up its independence.) However, the NATO countries were not willing to match the Soviet Union's huge and expensive conventional forces and decided to use nuclear weapons as a balance. Under continued pressure from NATO members, a large stockpile of nuclear bombs and missiles was stationed in Europe to counter the massive Soviet conventional forces. Debate continued about what authority, if any, the Europeans should be given in the use of such weapons, discussions that went under the descriptor of "dual key" authorization. The United States wanted the Europeans to think that they had some control over the use of nuclear weapons on their soil, but it was unwilling to simply hand over an atomic arsenal. Dual key authorization required the host country and the American owner to agree before using a nuclear weapon.

The closest the world ever came to nuclear Armageddon was the Cuban Missile Crisis in October 1962. Alarmed by a clear American lead in both the number and quality of its ballistic missiles, and frustrated at the recent placement of U.S. missiles in Turkey, Khrushchev reached agreement with the Castro government to put Russian missiles in Cuba. While only a relatively small number of weapons were involved, their proximity to the American mainland meant that all the major cities on the eastern seaboard could have been destroyed before America could mount a response. In a rare break from secrecy, the U.S. government showed overhead spy photos of Soviet installations in Cuba and suspicious cargoes on incoming ships in an attempt to force world opinion against the Soviets. Intense negotiations were conducted around the clock, some in the open and some in private phone calls between Washington and Moscow, ending in the agreement that the Soviets would withdraw their weapons from Cuba if the United States withdrew ours from Turkey. Emotions ran high on both sides. Whether the world averted nuclear catastrophe by diplomatic skill or simple luck is left for historians to debate.

LIMITATIONS ON OFFENSIVE weapons took a step forward with the Strategic Arms Limitation Talks (SALT) that began in November 1969. Both sides recognized that the number of weapons had grown out of proportion to their practical use in war fighting or even in maintaining a strategic balance. While the United States had an advantage in multiple independent reentry vehicle (MIRV) technology that allowed more than one warhead to be placed on a single missile, the Russians were greatly increasing the number of their ballistic missiles and were well on the way to their own MIRV technology. After intense negotiations, the United States and the Soviet Union agreed to limit the number of "strategic" ICBM silos and missiles on submarines. However, the treaty was silent about the many thousands of "tactical" nuclear weapons in the arsenals of both sides, an omission that has continued to complicate arms negotiations to the present day.

WORK ON DEFENSES against nuclear attack increased in the late 1960s, with pressure mounting to deploy a missile defense system in the United States. Lacking the technology to hit an incoming warhead, the Sentinel missile was tipped with a nuclear explosive that was intended to destroy the enemy weapon before it entered the atmosphere. However, owing to opposition to having nuclear-tipped interceptor missiles scattered throughout the country, President Nixon scaled the system down to the defense of a single ICBM field in Grand Forks, North Dakota. Placing defenses around the missiles meant that the Soviets would never be sure of eliminating all of them in a first strike, leaving some for use in an American counterstrike.

To prevent yet another escalation of the arms race that would pit expensive defensive systems against equally expensive offensive missiles, the United States and the Soviet Union signed the Anti–Ballistic Missile (ABM) Treaty of 1972. This agreement limited deployment of missile defenses to a maximum of two hundred defensive interceptor missiles defending two sites. (An amendment cut this to one hundred missiles defending a single site.) The United States chose to maintain its plan to defend Grand Forks, while the Soviets chose to deploy their interceptors around Moscow. However, the high cost and relative ineffectiveness of the American system led to its termination in 1976, after only a few years of operation. (The Soviet system continues in operation as I write this.) In a sense, the ABM Treaty cemented mutually assured destruction as a strategy for nuclear warfare in that it limited the ability of either side to defend itself against a massive attack. Vulnerability was set by treaty.

JAMES SCHLESINGER, secretary of defense during the Nixon administration, revived concerns about nuclear war plans that, despite the doctrine of "flexible response," were still focused on a massive response to Soviet nuclear attack. He reasoned that the extreme damage caused by a counterattack essentially "self-deterred" the United States from ever using nuclear weapons. No American president would order the deaths of tens of millions of civilians, no matter the provocation. To make deterrence believable, Schlesinger argued that "sub-SIOP" options should be developed for the president. The damage caused by such "small" options was still severe, since they involved hundreds of weapons, but it was considerably less than the civilization-threatening destruction that would result from a full-scale strategic exchange of thousands or even tens of thousands of nuclear weapons. Also, increasing precision in ballistic missiles meant that the notion of a "counterforce" strategy, one that targeted military sites rather than population centers, could again be considered, another step away from a blind application of mutually assured destruction. Schlesinger's approach of selected strikes told the Soviets that, if forced to use nuclear weapons, our response would be measured and well short of total war. Emphasis was placed on strategies for "war termination" on terms favorable to the United States and on increasing the number of options available should nuclear use be required. However, critics, who might have been rereading Dwight Eisenhower's notes, claimed that a limited response actually made nuclear war more likely by making damage more endurable, at least as seen from the cold calculations of military planners. Once again, the conundrum of use versus deterrence reared its head.

NEGOTIATIONS FOR A follow-up to the SALT Treaty were complicated by the rapid advance of technology, including the development of supersonic long-range bombers, cruise missiles, and the proliferation of tactical (i.e., short-range) missile systems in Europe and elsewhere. President Jimmy Carter, taking office in 1977, recommended dramatic reductions in the weapons stockpiles of both countries, a notion rejected by the Soviets, who insisted on keeping to the SALT agenda of moderate controls. Carter did cancel the development of the new long-range B1 bomber and ordered a new study of the proposed MX intercontinental ballistic missile. The latter system was designed to be mobile—the missiles were to be mounted on giant trucks or railcars so that the enemy could never be sure of locating and destroying all of them in a first strike. (President Reagan eventually deployed one hundred of them, with the name "Peacekeeper," but in fixed silos rather than mobile launchers.)

Carter reemphasized the policy of counterforce and diverted war planners' focus from Soviet industrial sites to nuclear forces, a further shift from a strategy designed to damage the Soviet Union to one that prevented it from achieving its war aims. He also announced a "no first use" policy for non-nuclear nations that promised not to develop nuclear weapons of their own. The aim was to convince them that there was an advantage to staying nonnuclear. After contentious discussions, the SALT II Treaty was completed in 1979, only to be shelved by the United States following the Soviet invasion of Afghanistan, an example of how nonnuclear events can scuttle treaties presumably in the best interests of both parties.

RONALD REAGAN TOOK office in 1981 with strong views on international relations. He saw the Soviet Union as the "Evil Empire" and thought that security could only be maintained by a manifest show of strength in all areas of military preparedness. However, Reagan, like all presidents since Truman, disliked the very idea of nuclear weapons and sought to find some new way out of the strategic morass of mutually assured destruction. Part of the solution, he reasoned, would be to engage the American scientific establishment in the development of a "leakproof shield" of ballistic missile defense that would nullify any advantage that the Soviets might have in attacking the United States. While certainly not the first attempt at large-scale defensive systems, the Strategic Defense Initiative (popularly called Star Wars) tried to shift the focus of nuclear strategy away from a massive counterattack toward defense against a first strike. At the same time, Reagan began a sweeping modernization of the American nuclear deterrent that included new bombs and warheads and a much more accurate missile for the navy that allowed precision targeting of Soviet military installations by sea-launched weapons.

While he was overseeing one of the largest arms buildups in recent history, Reagan met with Soviet leader Mikhail Gorbachev to discuss nuclear arms control measures. Fearing that the Soviets would break any agreement not to their liking, Reagan focused on intrusive inspection measures in accord with his mantra of "Trust, but verify." For the first time, American and Soviet inspectors would visit the nuclear installations of the other side and see with their own eyes that limits on warheads, bombers, and missiles were being observed. Not to be outdone, and perhaps recognizing that the Soviet Union could not hope to compete with the United States in Reagan's massive revamping of nuclear systems, Gorbachev went even further to propose the elimination of all nuclear weapons by the year 2000, on the condition that the Americans give up on missile defense. Reagan refused, convinced that defense offered an escape from the MAD doctrine of his predecessors. The option of nuclear disarmament, real or imagined, faded from the superpower agenda.

Arms control initiatives continued on other fronts with agreements to remove short-range nuclear missiles from Europe and to limit the energy of nuclear test explosions to no more than 150 kilotons. The latter agreement included the option for the other side to place monitoring equipment at the test site to verify that the limit was being observed.

THE END OF the Cold War presented new challenges to the nuclear strategist, most notably that both sides were left with massive nuclear arsenals without any apparent missions. With Russia no longer considered an adversary of the United States, what was the purpose of strategic nuclear weapons? What was the role of tactical nuclear weapons in Europe when the Soviet conventional threat had disappeared and German unification was now the topic of the day? President George H. W. Bush attempted to take advantage of the new world order by pressing forward with a new round of arms limitation talks that concluded with the START I Treaty of 1991. (For various reasons, not least of which was the collapse of the Soviet Union, the treaty did not take effect until the end of 1994.) He canceled several American nuclear weapons programs, took strategic bombers off alert, and ordered the destruction of all short-range nuclear weapons such as those in artillery shells and on short-range rockets. The Strategic Air Command, a symbol of the Cold War nuclear standoff, was disbanded and replaced with United States Strategic Command, an organization with a broader mandate that included defense as well as offense. Finally, Bush ordered the cessation of all nuclear tests, with the intention of limiting any future arms race and to discourage other countries from using U.S. testing as an excuse for their own programs.

Perhaps the most remarkable effort of the post–Cold War period was initiated in 1992 by two U.S. senators—Sam Nunn of Georgia and Dick Lugar of Indiana. In an unprecedented act of cooperation, one that would hardly have been dreamed of just a decade before, they created the Cooperative Threat Reduction (CTR) program that provided assistance to the Russians in destroying large numbers of redundant bombers, missiles, silos, submarines, and other relics of the Cold War. American advisers and equipment were sent to Russia to cut the wings off bombers, blow up concrete missile silos, and chop up nuclear submarines. In the years since its inception, the CTR program has destroyed hundreds of strategic weapons platforms, achieving more than any arms control program in history. Given the enormous cost of weapons development during the Cold War, it may have been the best use of defense dollars ever.

PRESIDENT BILL CLINTON, inaugurated in 1993, continued the disarmament momentum established by George H. W. Bush and attempted to make the test moratorium permanent by signing the Comprehensive Test Ban Treaty. However, after extensive discussions, the United States Senate was unconvinced that computer calculations and laboratory experiments could maintain the nuclear deterrent indefinitely, and it refused to ratify the treaty. While the United States has not conducted a nuclear explosion since 1992, it is under no legal obligation to refrain from testing should the need arise.

Other arms controls measures, most notably the START II Treaty of 1996, were negotiated to further limit long-range nuclear forces, but turmoil within the newly created Russian Federation led to Russian insistence on several provisions that were not accepted by the American side, leaving the treaty in limbo.

THE SECOND PRESIDENT BUSH'S term in office was dominated by the terrorist attacks of September 2001 and the subsequent wars in Afghanistan and Iraq. However, he took a keen interest in nuclear matters. The Treaty of Moscow, a remarkably short and informal document considering the complexity of its predecessors, mandated that the number of strategic warheads be reduced to the range of seventeen hundred to twenty-two hundred, a major reduction from Cold War levels. Bush recognized that the geopolitical situation had changed in a fundamental way and that he had the opportunity to significantly reduce a nuclear threat that might arise from a recidivist Russian government.

Part of Bush's willingness to reduce the American nuclear stockpile was based on the clear superiority of American conventional military hardware, a superiority graphically demonstrated in the two Persian Gulf wars. Russian generals were shocked when the Iraqis, equipped with reasonably modern Soviet military hardware and tactics, failed dismally in their response to American assaults on land and in the air. What had just a few years before appeared as a massive Soviet military juggernaut turned out to be a paper tiger. Bush capitalized on the American technological advantage and broadened the mission of the United States Strategic Command from being exclusively nuclear to include all aspects of strategic attack and defense, including long-range conventional weapons (such as conventional warheads on high-precision ballistic missiles), cyber warfare, and missile defense. For the first time since the invention of the atomic bomb, other technologies were seen as *superior* to nuclear weapons in achieving certain high-priority missions.

However, any gloating on the part of American generals was short-lived when, taking a page out of Cold War NATO strategy, Russia realized that the inferiority of *its* conventional forces could only be compensated for by a greater reliance on tactical nuclear systems. During the early years of the new century the Russians embarked on an extensive program of modernization of their nuclear forces, including new low-yield weapons on short-range missiles and maneuvering warheads that could evade American missile defenses.

OVER THE SIX decades since Hiroshima, nuclear strategy has oscillated between deterrence based on assured destruction and attempts to limit the destruction from any nuclear war that did occur. The end of the Cold War, the defining strategic standoff of the twentieth century, presents a new opportunity for creative thinking in dealing with nuclear issues, one that must not be squandered lest we find ourselves back in the same debates that challenged Truman, Eisenhower, and Kennedy.

3

Current Nuclear Arsenals

The United States has dismantled thousands of weapons following the end of the Cold War, but we still maintain a massive nuclear arsenal. To get an idea of just how many weapons we have, consider the following: The U.S. Air Force conducts periodic exercises on our fleet of B52 bombers to assure that they are ready for a nuclear strike. Aircraft are assembled at Barksdale Air Force Base in Louisiana and crews practice mounting bombs and missiles, maneuvering the planes into takeoff position, and other aspects of nuclear war. All this must be accomplished within a set period of time to assure that, should this capability ever be needed, it will be ready.

I observed one such exercise while I was a member of United States Strategic Command's Strategic Advisory Group, a body that advises the commander on various issues. In blustery, cool weather we watched enlisted men mount a brace of cruise missiles on the wing of a bomber and talked to the crew about their simulated mission. Later, we were driven along the flight line of aircraft that, parked wingtip to wingtip, stretched for more than a mile, a display of strategic might visible from Russian spy satellites. We passed plane after plane after plane, each capable of carrying many strategic nuclear bombs or nuclear-tipped cruise missiles. One could hardly imagine a more vivid demonstration that nuclear weapons are still very much a part of the defense equation. Match this with hundreds of land-based intercontinental ballistic missiles and their submarine equivalents, and one can understand that the American nuclear deterrent is potent indeed.

Many different concepts for nuclear explosives have been proposed and tested, but most modern strategic weapons are based on a two-stage design consisting of a primary and a secondary, as shown schematically below. The primary stage contains high explosive and plutonium and is similar in principle to the implosion bomb dropped on Nagasaki, although much smaller and lighter. However, the main purpose of the primary in a modern weapon is to generate enough energy to implode the secondary stage, which actually produces most of

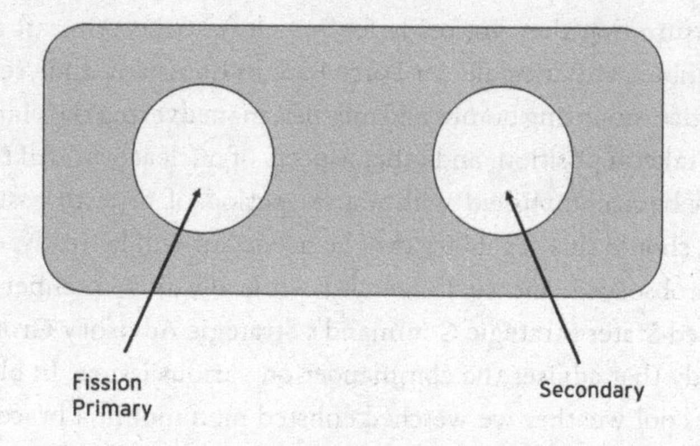

Schematic of a modern nuclear weapon. The purpose of the primary is to produce enough energy to implode the secondary, which generates most of the yield of the weapon.

the yield. The secondary is what makes a two-stage weapon a "hydrogen bomb," since it usually involves the fusion of isotopes of hydrogen in its operation. Smaller weapons might contain only a primary stage, dispensing with the secondary altogether.

Whereas the implosion bomb developed during the Manhattan Project weighed more than four tons and was nearly five feet in diameter, a modern primary weighs less than a few hundred pounds and is usually less than two feet in diameter. Hundreds of nuclear tests were required to refine the design to this level since, as with all highly optimized systems, there is a risk of failure when one cuts margins too thin. Every effort has been made to make the primary both reliable and safe—it must have a near perfect probability of working when needed and a probability of less than one in a million of accidentally exploding in the most severe accident. Some weapons employ "insensitive high explosive" (IHE) that is less susceptible to detonation should the weapon be subjected to fire (e.g., in an aircraft crash) or shock (e.g., bullets or fragments from anti-aircraft missiles).

Primitive nuclear weapons designs sometimes separate the plutonium from the rest of the weapon until just before use, an added safety feature when the amount of fissionable material is so great that an accidental nuclear detonation could occur in an accident. Security features have been built into modern weapons that make it impossible for an unauthorized person to detonate them.

To a first approximation, the larger the secondary, the more yield it will produce. Various types of advanced secondary designs have been investigated for special applications, but the degree of complexity associated with such advanced designs puts them beyond the reach of most entry-level nuclear powers and certainly beyond the range of proliferants who have never tested a nuclear explosive.

The primary and secondary are mounted in a case, which contains the energy generated by the primary long enough for the secondary to implode and produce its yield. Cases are typically made of heavy metals.

Security restrictions make detailed information on nuclear weapons hard to come by. While all the acknowledged nuclear weapons states have published some account of their forces, close inspection often reveals little detail beyond the fact that different types of weapons exist. The official government position is one of purposeful ambiguity, a refusal to officially confirm or deny what is otherwise generally known. The number of each type of weapon in the American arsenal is classified, as is its yield and geographic location. Several publications have attempted to collect basic data about nuclear weapons, but classification rules prohibit me from commenting on the accuracy of any of these estimates.

Before surveying the nuclear weapons in the arsenals of the major nuclear powers, it is worth clarifying some terminology. Considerable confusion has arisen from misunderstanding the different categories of weapons, how they are counted, and even how the energy of weapons is measured.

In the United States, nuclear weapons mounted on missiles (either long-range intercontinental ballistic missiles or shorter-range cruise missiles) are called *warheads* and are identified with a "W" and a numerical designator, as in the W78 warhead mounted on the Minuteman III ballistic missile. *Bombs*, or weapons that fall from aircraft, are given a "B" designator with a number, as in the B83 strategic bomb. Warheads and bombs were numbered in the sequence of their development—a weapon number is not the year in which it was deployed. The

W88 warhead for the Trident submarine-launched ballistic missile was the last nuclear weapon to enter the U.S. arsenal. Its design dates from the early 1980s. Work had started on the W89, which was intended for an anti-aircraft missile, in the late 1980s, but it was terminated when nuclear testing stopped in 1992.

Most numbers quoted for the nuclear stockpile refer only to strategic weapons—typically higher-yield (hundreds of kilotons) designs meant for long-range delivery. During the Cold War, a separate stockpile of lower-yield (tens of kilotons) tactical weapons was developed to stop attacking tank and infantry formations and to destroy ships at sea. Nuclear war planners see little distinction between tactical and strategic nuclear weapons—they view any use of a nuclear explosive as a strategic event of the first magnitude. But most arms control treaties deal only with strategic weapons, those that are carried on bombers, missiles, and submarines and hence are easy to count.

The explosive output or yield of a nuclear weapon is measured in kilotons or megatons. A kiloton of yield is the equivalent of one thousand tons of TNT, a conventional explosive. To put that in perspective, one thousand tons of TNT would be a stack of explosives the size of a small house. A megaton is equal to one million tons of TNT equivalent, more explosive power than was used in most wars. A low-yield nuclear weapon is usually meant to be one with a yield below about ten kilotons. Very low-yield weapons, sometimes called micronukes, have yields below one kiloton. Most strategic weapons have yields in the several hundred kiloton range. High yield is typically taken to mean weapons in the megaton class.

Nuclear weapons are not just beefed-up versions of non-nuclear bombs dropped in World War II, Vietnam, or Iraq—they are *vastly* more powerful and produce destruction on a completely different scale. The biggest piece of conventional weaponry—the Massive Ordnance Air Bomb—has only about ten tons of explosive energy. A small nuclear explosive with a yield of ten kilotons is thus *one thousand times more powerful* than the largest conventional bomb. But there is another important difference between conventional and nuclear weapons. Conventional explosives destroy targets by means of a blast wave, essentially a mechanical effect. Nuclear weapons produce blast waves plus intense heat and radiation in the form of x-rays, gamma rays, and neutrons. Dirt and other matter swept up in the nuclear fireball is transmuted and then deposited in the form of radioactive fallout, the effects of which may linger for years.

TO REACH TARGETS on the other side of the planet, long-range missiles fly into space before releasing their warheads. Once separated from the missile, the warheads fly on a ballistic trajectory without any internal guidance, just as a rock would fly after it is thrown. The warheads must then reenter the atmosphere like miniature spacecraft, and, like spacecraft, they must contend with the intense heat produced by atmospheric friction. Photographs of warheads taken during missile tests show their exteriors to be white-hot just before impact. Protecting sensitive equipment inside ballistic missile warheads is a major technological hurdle to any nuclear state and one that only a few have mastered.

U.S. ballistic missile warheads are all mounted in reentry bodies (navy terminology) or reentry vehicles (air force terminology), black cones about two feet in diameter at their base and about four feet long. Their black color derives from the use of carbon composite heat shield material, similar in function to what is used on the space shuttle. They typically weigh several hundred pounds.

Older Soviet weapons were mounted in blunt nose cones and were much larger and heavier than their American counterparts. Their accuracy on target was typically lower than modern, sharper angled cones, but their higher yield made up for poor delivery accuracy.

Despite advances in the precision delivery of conventional weapons, nuclear bombs have no internal guidance—they are released by the aircraft and fall by gravity to their target. The accuracy with which they can be placed on the ground is a complex mixture of the skill of the pilot, the altitude at which the bomb is released, the prevailing winds, and the flight characteristics of the bomb. Many bombs have a parachute to slow their descent and to give the bomber time to get away from the area before detonation. Some have timers to delay their explosion still longer—a particularly important feature for pilots engaged in slow, low-altitude drops of megaton-class weapons. Modern strategic bombs measure about one to two feet in diameter and are about ten feet long. They weigh upward of several hundred pounds, much of that weight comprising the bomb casing and parachute.

To complicate things still further, nuclear weapons are also referred to by the Mark numbers associated with the delivery packages in which they are housed. For example, the W88 weapon is mounted in the Mark 5 reentry body, and the W78 weapon is mounted in the Mark 12A reentry body. Sometimes the nuclear explosive part of a weapon is referred to as the "physics package" to distinguish it from the weapon's electronic subsystems, external housing, parachute (if any), and other associated equipment.

The nuclear weapons stockpile is divided into two parts.

The *active stockpile* consists of those weapons that are ready for immediate use. The *inactive stockpile* consists of weapons kept in lower states of readiness in secure warehouses. In addition to weapons mounted on missiles and those that are ready to be loaded into bombers, the active stockpile includes units that are in the logistics tail of the supply chain, such as those that are awaiting replacement on submarines currently undergoing a refit, those that are en route to new locations, and those undergoing periodic refurbishment.

The inactive stockpile contains weapons that do not receive the same degree of attention as active bombs and warheads, but that could potentially be returned to service. Some short-lived components, such as batteries, are removed from weapons in the inactive stockpile, and they receive fewer inspections. Weapons in the inactive stockpile are not countable under arms control agreements. The purpose of the inactive stockpile is to provide backups for active weapons should a catastrophic flaw be discovered. They also enable the stockpile to be rapidly expanded should geopolitical events worsen more quickly than new weapons can be manufactured. It is possible for the number of weapons in the inactive stockpile to exceed the number of officially recognized weapons, making weapons counting a topic of some confusion even for nuclear planners.

Nuclear Weapons Systems of the United States

Only the air force and navy have nuclear weapons—neither the army nor the marines currently has custody of any nuclear explosives. Also, only the air force has both bombs and missiles—the navy has only missile warheads. The table gives a current summary of U.S. nuclear forces.

The B61 bomb was first introduced into the stockpile in

1967 and has been modified many times since. Some are intended for tactical battlefield applications and some for strategic uses. Having these choices available in a single type bomb is a considerable cost savings for the air force, since only one type of aircraft mounting is required for multiple missions. Also, training requirements are reduced as airmen are not forced to be proficient on many different weapons systems. As shown in the photograph on page 78, the B61 consists of hundreds of individual parts, most of them associated with the electronics and other systems that arm and fire the weapon.

Nuclear weapons currently deployed by the United States. Here, ICBM (intercontinental ballistic missile) refers to land-based missiles and SLBM (submarine-launched ballistic missile) to submarine-based missiles.

WEAPON	TYPE	DELIVERY VEHICLE
B61	Bomb	B52 and B2 heavy bombers and several fighter bombers
B83	Bomb	B52 and B2 heavy bombers
W76	SLBM Warhead	D5 missile on Trident submarine
W78	ICBM Warhead	Minuteman III ICBM
W80	Cruise Missile Warhead	Cruise missile carried by bombers and submarines
W87	ICBM Warhead	Minuteman III ICBM
W88	SLBM Warhead	D5 missile on Trident submarine

The majority of the bombs in the U.S. arsenal are variants of the B61 family, the most recent one being the B61 Mod 11 earth-penetrating bomb, or B61–11, which was adapted from an earlier B61 by adding a thicker case and a stronger nose cone. These changes enable it to penetrate a few yards into the ground after being dropped from an aircraft, greatly increasing the amount of energy sent into the earth to destroy deeply buried targets such as command centers or weapons storage areas. The B61–11 has the destructive power of a surface burst many times larger.

I oversaw the engineering testing of the B61–11 and transferred it from the research and development phase at Los Alamos to service in United States Strategic Command. While the basic design of the nuclear explosive was well verified via underground nuclear tests, considerable work had to be done to verify that it could survive hitting the ground at high speed and penetrate to the required depth. Imagine dropping a highly sophisticated piece of equipment from a great height and expecting it to perform flawlessly after impact. There was considerable concern that this long, thin cylinder might break in the middle or that delicate components might be damaged by the shock of impact. More than that, we had to contend with the possibility that the weapon might land on concrete or hit a rock during its entry into the ground, and a myriad of other possi-

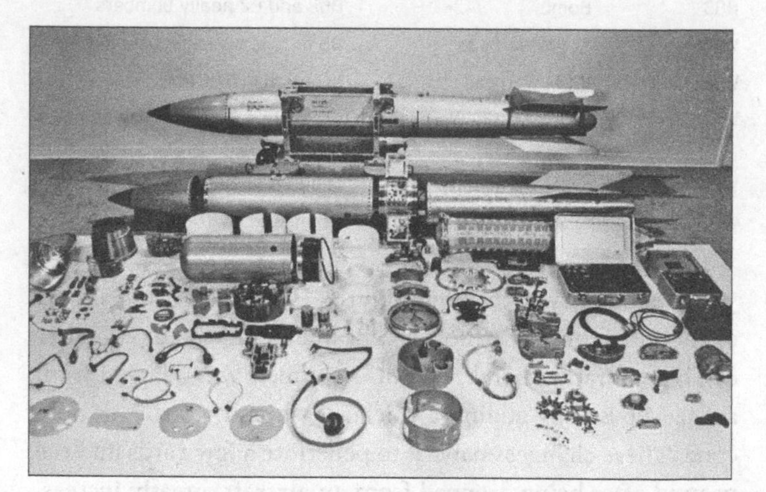

The B61, the most common nuclear bomb in the U.S. stockpile, with its interior parts spread out in front. The "physics package," or nuclear explosive, is the cylinder in the center left.

bilities for which the basic B61 was never intended, all with a design that was approaching forty years old.

The other bomb in the U.S. stockpile is the B83, a megaton-class strategic bomb. Since it is a much more modern design, the B83 has a variety of features not present in the B61, making it inherently safer and more reliable. However, the increasing precision with which even unguided bombs can be delivered means that there are few targets that require such a large yield for their destruction.

The W80 warhead carried on bomber- and submarine-launched cruise missiles was designed as a standoff weapon that could be launched when a bomber was far from the dangers of Soviet air defense systems. Once released, the missile follows a preset flight plan, sometimes flying quite low to avoid

Photo of a B61-11 earth-penetrating weapon falling from a B2 bomber.

detection by enemy radar. It thus combines the advantages of the bomber—the ability to cancel a mission before the weapon is dropped—and an unmanned missile that does not expose a crew to dangerous flight over enemy territory.

The air force maintains two warheads for use on intercontinental ballistic missiles: the W78 and the W87. (A third warhead, the W62, was recently withdrawn from the stockpile.) Having more than one warhead for ballistic missiles reduces the effect of a single-point failure in any class of weapon, a serious problem given the long time required (several years) to fix hundreds of units or manufacture entirely new ones. For ex-

The author next to the W78 warhead that is carried by the Minuteman III intercontinental ballistic missile.

The air-launched cruise missile (ALCM), which can be carried on a heavy bomber. It is capable of carrying a conventional or nuclear warhead.

ample, if it was found that all the W78 warheads suffered from a serious problem related to their age, the W87 would still be available to maintain deterrence. The photograph on page 80 illustrates a typical ICBM warhead, in this case the W78, which flies on the Minuteman III missile. The W87 looks quite similar but includes a number of modern safety features not present in the older weapon.

Bombs and warheads are only one part of a nuclear weapons *system*, which includes the explosive package, its delivery vehicle (bomber, missile, or submarine), and the command and control system that governs its potential use. The air force relies on two heavy bombers—the B52 and the B2—to carry its strategic bombs and air-launched cruise missiles. The B52 was originally introduced in 1954 and has undergone significant upgrades since that time, including the installation of new wings,

The nuclear warhead carried on the ALCM.

more efficient engines, and greatly improved electronics. The air force currently plans to keep these workhorses well into the present century, making it possible for five generations of pilots to fly the same type of airplane. But age alone is not the sole criterion for reliability—the average B52 in the fleet has less flight time than the newer 757 commercial passenger airplane, which receives much more frequent use. It is quite possible for a properly maintained aircraft to keep flying safely for many decades.

Since a bomber must get close to a target (in the case of cruise missiles) or actually on top of it (in the case of bombs) to deliver weapons, it is especially vulnerable to fighter aircraft and anti-aircraft missiles. The B2 stealth bomber was a significant advance in large aircraft design, with numerous features that make it exceptionally difficult to detect on radar or by other means.

The venerable B52 heavy bomber, which first flew in 1954, is still assigned to conventional and nuclear missions.

The air force has several kinds of fighter-type aircraft that can carry its tactical bombs. These have much shorter flying ranges than heavy bombers and were originally intended for use in a war between the Soviet Union and NATO forces in Western Europe.

The Minuteman III missile, originally deployed in 1970, is slated to carry air force warheads well into the twenty-first century. Like most long-range missiles, it has multiple stages—in this case three—that successively fall away as the missile ascends. It is powered by solid fuel, so that it can sit in a silo for years or even decades with relatively little maintenance. The Minuteman III has a range of eight thousand miles and can deliver its warheads with remarkable accuracy—after flying from the continental United States to a target on the other side of the globe, it can place its weapons within a city block.

During the Cold War, a single missile sometimes carried more than one warhead, the MIRV (multiple independently targetable reentry vehicle) concept, but under the guidelines of

The B2 stealth bomber is nearly invisible to enemy radar and is capable of carrying nuclear weapons.

the Moscow Treaty of 2002, only one warhead will be mounted on a missile. This is consistent with a shift away from massive retaliation and toward the use of one or a few weapons to destroy very high-value targets in an extreme emergency. The Minuteman III is launched from hardened concrete silos located at several bases in the central part of the country.

The air force formerly maintained a fleet of Peacekeeper (or MX) missiles that, while more modern than the Minuteman III, were designed to carry more warheads than is currently required in operational plans. Also, the cost of maintaining the Peacekeeper was higher than that of the Minuteman III, so some savings were achieved by its removal.

The navy has two types of ballistic missile warheads, the W76 and the more powerful W88, both of which are mounted on the Trident D5 submarine-launched ballistic missile. Each Trident missile is capable of carrying multiple warheads and

Technicians work on the warheads of a Minuteman III intercontinental ballistic missile at Malmstrom Air Force Base, Montana.

there are no plans to reduce that number to one, as in the case of air force ballistic missiles. Like the Minuteman III ICBM, the D5 missile is a three-stage, solid fuel design, although its dimensions—forty-two feet long and six and a half feet in diameter—are tightly constrained by the necessity of mounting it vertically within a submarine hull. It has a maximum range of seventy-five hundred miles, quite impressive given that it can be launched from anywhere in the world's oceans.

The United States currently has fourteen ballistic missile

USS *Pennsylvania* (SSBN-735) ballistic missile submarine under way.

submarines (naval designation SSBN), members of the Ohio class of ships named after the first of that type introduced in 1990. Only twelve of these are on active duty at any given time, the remaining two undergoing refurbishment. Each submarine carries twenty-four Trident D5 missiles, giving a single SSBN the ability to project an explosive force greater than all the weapons used in all the wars of history—they are the most destructive weapons systems ever created by humankind. Because ballistic missile submarines are nuclear-powered, they can remain submerged for months at a time and are exceptionally difficult to detect with even the most advanced sonar technology. The commander of the submarine has some freedom to set his own course while on patrol, so only a few people, all of whom are on the submarine itself, know its exact location at any given time. Sophisticated communications systems allow the ship to remain in constant contact with the United States so it is always available as part of the overall deterrent force.

A Trident D5 missile shortly after launch from a submerged ballistic missile submarine.

The navy maintains cruise missiles equipped with a variant of the W80 nuclear warhead, the only case where an almost identical warhead is used by the air force and the navy. These missiles can be launched from attack submarines, giving increased flexibility to the nuclear war planner.

TO FULLY UNDERSTAND how the nuclear deterrent functions, it is important to understand the command and control systems that govern weapons use. Enormous effort has been expended

A member of the 321st Strategic Missile Wing works at a control panel inside a Minuteman III intercontinental ballistic missile silo at Grand Forks Air Force Base, North Dakota.

to ensure that U.S. nuclear weapons can be used only upon receipt of an authenticated presidential order. Codes, which are for all practical purposes unbreakable by any current or anticipated computer or mathematical technique, are carried in safes aboard submarines and bombers and kept in the control bunkers of ICBM sites. Upon receipt of a launch order, the code associated with the order is compared with the code stored in the safe—only when they agree can a launch proceed. As an additional safety precaution, modern weapons have various interlocks that are intended to prevent their detonation until a fixed sequence of events has occurred. For example, internal sensors on a weapon might verify that it has endured the acceleration of missile launch, a minimum required flight time, and the deceleration associated with reentry into the atmosphere, and that it is within a fixed set of altitudes suitable for detonation.

Nuclear Weapons Systems of the Russian Federation

The Russian Federation inherited the nuclear arsenal of the Soviet Union, including all the weapons that were formerly stationed in the Ukraine, Belorussia, and Kazakhstan. Like the United States, Russia maintains a diverse set of nuclear weapons, ranging from low-yield warheads intended for use against a ground invasion to high-yield strategic weapons for use against other countries.

Russia is in the process of modernizing its nuclear forces with the deployment of the SS–27 intercontinental ballistic missile, first observed in 1994. While the plan was to deploy one regiment of SS–27s—ten missiles—per year, economic difficulties in the country have delayed that schedule so full capability is not expected before 2010. The Topol-M RS–12M1, as the SS–27 is called in Russia, is a three-stage solid propellant missile that is seventy-two feet long and six and a half feet in diameter. According to the military journal *Jane's Strategic Weapons Systems*, the Topol-M has a range of sixty-five hundred miles and an estimated accuracy of eleven hundred feet. It can be launched either from underground silos or from an over-the-road transporter-erector-launcher (TEL). President Vladimir Putin announced in 2007 that the warheads on the SS–27 were capable of maneuvering around U.S. interceptor missile warheads, enabling them to penetrate our missile defense system.

Russia has a number of other nuclear-capable missiles, including the SS–25 Sickle design, which can be carried on a mobile trucklike launcher, and the SS–19 Stiletto silo-based missile. The massive SS–18 Satan missile is reported to be fitted with a multimegaton nuclear warhead, perhaps the most powerful explosive in the world—much more destructive than anything in the U.S. arsenal. At the other end of the spectrum, Russia has very capable short-range missile systems, some of which may be equipped with nuclear warheads. The purpose of these weapons is to provide a defensive capability during a period of shortfalls in Russia's conventional forces. If one is forced to use a nuclear weapon on one's own soil, it is wise to use the lowest yield possible.

Russia maintains a fleet of ballistic missile submarines, including the huge Typhoon submarine and the workhorse Delta class. Russia is introducing a new ballistic missile submarine, the Borey, which is intended to carry a new type of missile. The first of these new submarines was launched in 2007. Funding shortages within the Russian navy have prevented it from deploying its submarines at anywhere near the rate that it did during the Cold War, but this appears to be changing as the country stabilizes.

Finally, Russia is modernizing its strategic bomber force with improvements in the number and capabilities of its Blackjack heavy bombers. Just as the United States is maintaining our B52s, the Russians may elect to retain some of their Cold War fleet, in their case, the venerable Bear-class long-range aircraft.

Russia has a very active research and development program to improve its nuclear arsenal. It is designing new nuclear weapons for use on short-range missiles and it is conducting an extensive program of experiments at its nuclear test site at Novaya Zemlya, north of the Arctic Circle. President Putin repeatedly stated the vital role that nuclear weapons play in Russian national security and his intention to deploy new types of weapons.

Nuclear Weapons Development by Other Nations

The United States, being the first nuclear power, had to solve all the problems of atomic weapons on its own, a mammoth task involving the greatest concentration of scientific talent the world had ever seen. The Soviet Union had the advantage of U.S. data obtained by espionage, although it reproduced much of that work out of fear of being led astray by American counterintelligence. The United Kingdom worked closely with the United States during the Manhattan Project and had inside knowledge of how a weapon worked. However, British access to nuclear weapons information was suspended over espionage concerns—it was not until an agreement was signed in 1958 that cooperation was reestablished on a controlled set of topics. Hence both the Russian and British weapons programs were, in different ways, an offshoot of the American program.

France knew that nuclear weapons were possible when it set out to develop an independent nuclear deterrent. However, French scientists did not have access to data from successful nuclear powers, so they were forced to reinvent much of the technology on their own. (It was only later that the United States and the United Kingdom created carefully controlled interactions with France on nuclear matters.) France apparently went down a number of blind alleys in nuclear tests done in the Sahara Desert during the early 1960s, even though it presumably had its own espionage program aimed at America and the Soviet Union. Later experiments done at France's nuclear test site in French Polynesia demonstrated that it has mastered the art of nuclear weapons design.

China received material help from the Soviet Union during the 1950s, an input that was canceled just prior to providing the Chinese with the design for a simple atomic bomb. The difficult step of turning concepts into functioning machinery had to be tackled without Soviet help, as did the completion of a working design and the means to test the performance of the device. How much assistance, if any, the Chinese had in the later development of their thermonuclear arsenal is a matter of conjecture, but their remarkable rate of progress in going from a simple atomic bomb to megaton-class warheads suggests that they had some help, either through voluntary sharing or involuntary espionage.

More recent entrants to the nuclear club had two advantages. First, much information about the basic operating principles of nuclear weapons exists in the open literature. Second, the overall level of science in the world has advanced tremendously since the days of the Manhattan Project. Topics like computational hydrodynamics and opacity theory, which were invented in part to develop the implosion design during World War II, are now taught in nearly every university. Even though the application of this information to nuclear weapons is neither confirmed nor denied by the governments of nuclear powers, scientists in other countries can draw their own conclusions, and they increasingly have the tools necessary to turn theory into real weapons.

NUCLEAR FORCES OF THE UNITED KINGDOM

The British government summarized its future nuclear policy in a 2006 white paper: "We are committed to retaining the minimum nuclear deterrent capability necessary to provide effective deterrence, whilst setting an example where possible by reducing our nuclear capabilities, and working multilaterally for nuclear disarmament and to counter nuclear proliferation. We believe this is the right balance between our commitment to a world in which there is no place for nuclear weapons and our responsibilities to protect the current and future citizens of the UK."

The nuclear forces of the United Kingdom consist of four ballistic missile submarines, each fitted with sixteen Trident D5 missiles purchased from the United States. Given the need for periodic maintenance, crew changes, and other logistical tasks, this guarantees that the UK has at least one submarine ready to launch at any given time. In an emergency, it could count on two or even three. The UK has a relatively small stockpile of warheads, numbering only about 200, and each submarine will likely carry fewer than this number while on patrol. Under their new plan, the British will reduce their stockpile to approximately 160 warheads carried on three or four new submarines. While the United Kingdom has benefited from a long collaboration with the United States in nuclear weapons technology, it maintains its own warhead design and manufacturing capability.

NUCLEAR FORCES OF FRANCE

At the dedication of the ballistic missile submarine *le Terrible* on March 21, 2008, President Nicolas Sarkozy announced that "our arsenal will include fewer than 300 nuclear warheads. That is half of the maximum number of warheads we had during the Cold War. In giving this information, France is completely transparent because it has no other weapons beside those in its operational stockpile."

France has four nuclear-powered ballistic missile submarines, each carrying sixteen M45 ballistic missiles with a range of thirty-seven hundred miles. These missiles are due to be replaced starting in 2010 by the M51, a missile that will have a range of six thousand miles. Both missiles will carry the advanced TN-75 warhead, although other weapons are planned for the future. France maintains sixty ASMP air-to-surface missiles fitted with the three-hundred-kiloton TN 80/81 warhead. The French nuclear weapons program has been carefully planned to ensure operational capability for decades to come.

NUCLEAR FORCES OF CHINA

China has only a few dozen ballistic missiles capable of reaching the United States, each of which is equipped with one or more nuclear warheads in the megaton range. The relatively high yield of China's nuclear warheads is required due to the poor accuracy of its missiles as well as the need to compensate for small numbers. To increase their survivability during times of war, Chinese missiles are launched from road-mobile TELs that can be rapidly dispatched from storage facilities. China has announced plans to modernize its nuclear forces, including land-based missiles and its own ballistic missile submarine.

NUCLEAR FORCES OF INDIA

India tested what it referred to as a peaceful nuclear explosive in 1974. Photographs suggest that the system was too bulky and heavy for deployment on aircraft and missiles. India claimed to have made significant advances in its nuclear test program of 1998, although there is some debate about the accuracy of those claims. The indigenously developed Agni missile, which has a range of twelve hundred to twenty-one hundred miles, may be fitted with a nuclear warhead, and Indian fighter-bomber aircraft could carry nuclear bombs. India is developing a ballistic missile submarine, but the government insists that this vessel is intended to carry only conventionally armed missiles and not nuclear weapons.

NUCLEAR FORCES OF PAKISTAN

Immediately following the Indian nuclear tests of 1998, Pakistan tested a nuclear device with an estimated yield of about thirty-five kilotons. It was reported in the press that this was done with some assistance from China. Pakistan has several short- and intermediate-range missiles that could carry a nuclear warhead, in addition to fighter-bomber aircraft that could carry nuclear bombs. A significant challenge for Pakistan (as well as India) is the development of an effective command and control system that will preclude the unauthorized use of nuclear weapons.

NORTH KOREA

North Korea reportedly conducted a nuclear test in 2006, demonstrating at least some capability in nuclear weapons technology. While hailing its success as indigenous, press reports suggest that help might have been obtained from other countries. However, the small yield of the North Korean test suggests that all may not have gone according to plan, either due to faulty information received from others, a misunderstanding of what was provided, or difficulties in the fabrication of the device.

IRAN

Iran is pursuing the enrichment of uranium to levels higher than would be required for a nuclear energy program. There is little public information available about Iranian progress on the other phases of nuclear weapons development, especially those related to mastering the design of an implosion device or the engineering associated with turning an experimental device into a practical weapon.

OTHER NATIONS

While the Israeli government has refused to confirm or deny that it has nuclear weapons, it is widely believed to have some capability. Several other countries appear to have had, or currently have, programs aimed at developing nuclear weapons. However, aside from claims and denials, there is often little evidence to support these assertions.

Nuclear status can change with the political environment, as when the newly independent countries of the former Soviet Union elected to return all nuclear weapons on their soil to Russia. The government of South Africa decided in 1990 to destroy both its nuclear weapons and the industrial plants that produced them. Other countries, such as Switzerland and Sweden, abandoned their programs in the 1960s.

LIKE AUTOMOBILES AND airplanes, nuclear weapons differ in design according to the missions that they are intended to perform and the technological capability of their builders. Just as one chooses the vehicle according to the task—racing or hauling bricks—a nuclear state plans its nuclear forces according to the types of missions they are intended to perform. Both the United States and the Soviet Union produced miniature nuclear explosives mounted in artillery rounds, weapons that were intended to stop massive troop and tank movements on the battlefield. Nuclear-tipped torpedoes were fielded, including some that flew on short-range rockets before entering the water near their target. Weapons of greatly different yield—from tons to megatons—were developed for different types of targets, the lower-yield weapons typically intended for smaller and softer targets and the higher-yield weapons for hard targets such as underground bunkers, or large targets such as airfields or naval bases.

Testing was vital to the exploration of nuclear weapons technology. The United States and Russia each conducted about one thousand such tests over a period of more than forty years. While the two countries have comparable scientific knowledge, they followed different design philosophies in the development of their stockpiles. The Americans chose very sophisticated designs that saved on missile costs, and the Russians chose rugged construction and ease of maintenance. Despite these technical differences, each country has an assured capability to project overwhelming force anywhere in the world. The cessation of nuclear testing in 1992 has severely limited, but not totally precluded, the development of new classes of nuclear weapons. Advanced nuclear states can adapt older, already tested nuclear systems to different missions, and small modifications can be made to existing weapons without the need to test. Weapons designers still worry that an accumulation of small changes could eventually reduce confidence in the safety and performance of weapons that are not tested.

The United Kingdom, France, and China conducted far fewer nuclear tests than did the United States and Russia, so one might assume that they have a correspondingly smaller range of tested designs for possible future deployment. In the case of India and Pakistan, each of which conducted at best a few tests, one can say only that they have the ability to produce a nuclear explosion. The status of their operational nuclear capability remains in question. The same could be said of North Korea. The nuclear capability of countries that have not conducted any tests is uncertain.

4

Targets and Targeting

On the morning of August 6, 1945, Hiroko Fukada was holding her baby in a crowded Hiroshima streetcar. It was hard to get transport in the early morning and the car was packed full. All of a sudden there was a flash, and an explosion shattered the windows of the car. Hiroko was cut by the flying glass, but she was not seriously injured. Many of those around her were slumped down dead. She looked down at her son, and saw that his head had been punctured by a piece of flying glass. He smiled up at her bloody face. She would always remember the smile. He died later that day.

Taeko Teramae was waiting outside her school when she saw a shiny object descending through the sky. She thought that it was very pretty. There was a flash and a bang and she was thrown to the ground. A slimy grit filled her mouth and she threw up. In a few moments she stood and realized that she did not seem to be hurt. But many around her were staggering around in tattered and burned clothing. Much later, she found a piece of mirror and held it to her face. Her eyes were swollen, her skin red. She was afraid that she had become a monster.

Kinue Tomoyasu saw her daughter off to work and went back to bed. There was an air-raid warning, and after the all-clear sounded, she got up, put away her bedding, and went to the window to look outside. There was a flash, and she found herself on the opposite side of the room. The window had blown out and glass was everywhere. From the direction of the flash, Kinue knew that the explosion came from the center of the city, near where her daughter worked. She dressed, swept up the glass, and set out to find her daughter. As she came into the city center, she passed hundreds of injured people, and many dead bodies. She searched and searched until a neighbor told her that her daughter was down by the riverbank. The pretty young woman was a mass of burns, and maggots were already in her wounds. It was too painful to pick them out. The mother held the daughter in her lap, calming her and telling her to hold on. "I don't want to die," the young woman cried. She lasted another nine hours. Going home, Kinue Tomoyasu was caught in a black rain. Sometime later her hair began to fall out and purple blotches appeared on her skin. She lived to old age.

The stories of the survivors of Hiroshima, taken from atomicarchive.com, have a sad similarity. There were the three colors: the yellow flash, the redness of the fires, the black rain bringing deadly fallout. Time itself seemed to skip several seconds—people remember standing by a window or in the street, and the next thing they knew they were lying down many yards away. And the noises. A bang from the blast, the crackle of flames, the rumble of collapsing buildings. All around, the injured called, "Mother! Mother!"—something that people sometimes do when they know that they are going to die.

THERE IS RELUCTANCE by some to use the word "target" in discussions of nuclear weapons. We removed targeting data from missile guidance computers at the end of the Cold War, a confidence-building measure with Russia, even though both sides knew it took only a few moments to reload the data. "Targeting" is taken by some to be equivalent to saber rattling, a needless provocation. But nuclear weapons are not ethereal objects of international diplomacy—they are capable of inflicting enormous damage. Within each target circle of the nuclear war planner's maps are hundreds of thousands of Hirokos, Taekos, and Kinues.

There are two approaches to nuclear targeting: *countervalue* and *counterforce*. Countervalue targeting aims to destroy cities, populations, and other things of value so as to shock the enemy into ceasing hostilities. It is typically employed by countries that have several hundred nuclear weapons (or fewer) and a limited number of missiles or aircraft with which to deliver them. Although few governments advertise that their nuclear weapons are aimed at heavily populated cities, most are forced into this position by the lack of accuracy of their missiles, which makes them ineffective against hardened military targets.

Counterforce targeting, on the other hand, aims to eliminate the adversary's capability to inflict further damage on your side. It involves targeting military bases, missile fields, submarines, and other targets of strategic value, and it tries as much as possible to avoid damage to population centers. In the case of the Soviet Union, many key assets of military command and control were located in and around the Moscow region, so such distinctions were more technical than practical—large numbers of people would have been killed in either strategy.

In any type of armed conflict, the type of weapon chosen for a particular mission depends on the target that is being attacked. For example, if the target is a single individual at close range, a pistol will suffice, but if the target is a tank, then pistol shots would merely bounce off, and an armor-piercing round is required. If the target is a tunnel complex deep underground, then conventional high explosives will have no effect, and the only means of destroying it is with a nuclear weapon.

We divide likely targets for nuclear weapons into four categories:

- Soft point targets, such as mobile missile launchers that might contain weapons of mass destruction. These targets are easily destroyed with minimal explosive force, but they may be difficult to find against the clutter of other vehicles, buildings, forests, and so forth.
- Soft area targets, such as air bases, army posts, and naval shipyards. These targets can be destroyed by conventional explosives, but they are so large that huge quantities would be required.
- Hard point targets, such as missile silos and command and control bunkers. Targets in this category are designed to withstand almost any attack with conventional high explosives.
- Super-hard targets, such as facilities buried under mountains. Such targets include command and control facilities, manufacturing sites for weapons of mass destruction, and safe refuges for government leaders. When a facility is buried under a thousand feet or more of hard rock, even nuclear weapons might not cause significant damage.

Before going into detail on each type of target, we need to discuss how nuclear weapons damage their targets and how much damage might be expected for a given weapon yield. Whereas a conventional bomb or artillery shell causes damage by means of a blast wave or shrapnel generated during the explosion, most of the energy of a nuclear weapon—about 80 percent—is released in the form of x-rays. The remaining 20 percent of the energy of the bomb is emitted in neutrons, gamma rays, and only a small fraction in the form of "hydrodynamic" or blast energy. The intense blast waves seen in films of nuclear explosions are actually produced when x-rays, neutrons, and gamma rays are absorbed in the air or ground.

One of the distinguishing characteristics of a nuclear weapon is the radioactive fallout generated by the nuclear reactions in the explosion. The effects of a conventional weapon are over with the explosion itself, but the effect of a nuclear weapon can linger for months, years, or even decades. During the fission process—the means by which most of the energy of a weapon is generated—uranium and plutonium nuclei break into pieces. These smaller nuclei are often radioactive in their own right, emitting damaging radiation over time. When a weapon is detonated at high altitude, above where the nuclear fireball would touch the ground, a weapon's radioactive fallout is limited to that produced by its own materials and whatever elements in the atmosphere might be "activated" by the bomb's radiation. Since this is a relatively small amount of material, it is quickly dispersed by winds and spread over a wide area, producing some long-term health effects but few immediate casualties from fallout.

A much bigger problem occurs when the weapon is detonated at the surface of the earth or at low altitude where large amounts of soil are swept up by the blast wave and activated by the lingering nuclear reactions in the mushroom cloud. This

No surface material
in Fireball

Mushroom cloud of debris
swept up from surface

Electromagnetic
pulse

Surface Detonation **High-altitude Detonation**

Comparison of low- and high-altitude nuclear explosions. A low-altitude explosion sweeps up a large quantity of soil, which becomes radioactive in the nuclear fireball. Winds may later deposit it as fallout. A high-altitude explosion produces much less fallout since it does not touch the ground, but it produces an electromagnetic pulse that could disable sensitive electronics over a wide area. The radiation from a high-altitude explosion can also interfere with satellites in orbit.

is illustrated in the figure above. Not only are more atoms activated, but more *types* of atoms are affected, atoms that can produce long-term radioactive effects on people and the environment. Curiously, it is not the most radioactive elements that are the biggest problem, since they decay relatively quickly. The greater concern is with radioactive nuclei that take years or decades to decay, making habitation of the target zone unhealthy for prolonged periods. However, it is incorrect to assume that a nuclear explosion will render an area forever uninhabitable. Both Hiroshima and Nagasaki are thriving cities today, and even the Nevada Test Site, the location of one hundred atmospheric test explosions, has recovered to the point where it is difficult to tell where the tests occurred. Only the twisted wreckage of test equipment is a clue to the enormous energy released. (The situation is different in the Marshall Islands, some of which were used for high-yield atmospheric nuclear testing. Low levels of radioactivity in the soil are concentrated by growing plants and trees, contaminating potential food sources used by the residents.)

While the effects of residual radiation may be subtle, they are not entirely absent. When scientists wanted to construct an experiment to measure very low levels of radiation from natural sources, they found that their results were spoiled by nuclear test contamination in the steel in their apparatus. Searching the world for a pure enough material, they found the rusting hulks of World War I German battleships scuttled at Scapa Flow in the north of Scotland. These ships were constructed before nuclear testing, and with the protection of the ocean, they were unsullied by radioactive fallout.

Nuclear weapons produce other effects beyond blast and radiation. The intense heat resulting from a nuclear explosion can start fires over large areas in a city or a forest and cause severe burns on people. Military planners refer to this as a "secondary effect" and often do not consider it in their calculations of damage. The flash of light from the explosion can cause temporary or permanent blindness. Observers of early atomic tests conducted in the atmosphere wore thick welders' goggles to avoid eye damage as they watched the flash and ensuing mushroom cloud.

When a nuclear explosive is detonated at very high altitude, above several tens of miles, an electromagnetic pulse is produced, directed toward the ground. The mechanism is simple: Gamma rays from a nuclear explosion travel downward and collide with atoms in the atmosphere. During these collisions the gamma rays knock an electron from the atom, producing an electrical current, much like the electrical current in a radio-transmitting antenna. And, just like the current in an antenna, the current generated by the gamma rays produces a radio signal. This radio signal can disrupt or even destroy sensitive electronic equipment such as computers and communication systems.

Since an electromagnetic pulse (EMP) is generated in the upper atmosphere and projected downward, the affected area from such an attack is quite large, potentially encompassing an entire country. However, the consequences of such an attack are limited to electronics (people and structures are not affected), and there is considerable debate within the scientific community about the sensitivity of electronics to a given type of pulse. Contrary to media reports, it is not true that an EMP attack from a typical strategic weapon would completely shut down the electronics within a country. First, the effect is statistical in nature—some systems will not notice the pulse at all while identical counterparts will be affected. Second, the most likely effect from an EMP attack is "upset" rather than destruction, that is, a temporary scrambling of the memory of a computer or the frequency of a communication device, something that is easily corrected by rebooting or resetting the device. (Upset can, however, have catastrophic consequences if the computer is the flight controller of an aircraft or another time-critical system.) Third, the EMP output from a typical device is degraded by several design issues so that few, if any, weapons currently deployed in military stockpiles will produce the maximum possible effect. Of all the nuclear effects, EMP seems the most prone to misunderstanding and misinterpretation.

Yet another effect can happen when a nuclear explosion

occurs in low earth orbit where many communication satellites operate. The explosion of the weapon produces a burst of electrons, which are captured in the magnetic field of the earth, greatly contributing to the naturally occurring radiation pattern that satellites must endure. Even a relatively small nuclear weapon—one with a yield of ten kilotons—could produce enough radiation to destroy many satellites, interrupting vital communications around the world.

The area that is affected by a nuclear weapon depends on the type of nuclear device used, its yield, and the hardness of the target. While most nuclear weapons in military stockpiles produce the greatest fraction of their energy in the form of x-rays, some are optimized for generating intense bursts of neutrons, so-called neutron bombs. These weapons have a smaller blast effect, but their neutrons are lethal to people, animals, and other biological organisms. They are designed for use against massive tank formations, since they kill the tank crews with less damage to the surrounding countryside.

The yield required for a nuclear mission, and the precision with which the weapon must be delivered, depends on the hardness of the target. Suppose that the target is a missile silo with a door constructed of ten feet of reinforced concrete. A few tons of nuclear yield—a so-called micronuke—might not cause enough damage to destroy the cover, meaning that the enemy could still fire its missile. A ten-kiloton nuclear explosion would need to be on top of the silo to assure its destruction, but a one-hundred-kiloton explosion could be a few hundred feet away. The yield required for destruction does not simply scale with distance—the energy of an explosion is sent in all directions, so the effect decreases as the inverse cube of the distance; to have the same effect twice as far away, an explosive has to be eight times as powerful. This is an illustration of the importance of accuracy—more accurate missiles allow much smaller explosives to be used. We have seen this in conventional warfare when a single precision bomb has destroyed targets that formerly would have required massive heavy bombing with destruction over a wide area.

If the desired outcome is to cause massive destruction to a soft target such as a city, one would detonate a nuclear explosion at an "optimum height of burst" to allow the shock wave and fire-starting potential of the weapon to impact the largest area. Detonating at too low an altitude would cause most of the weapon's energy to be diverted into making a crater, while too high a burst would dissipate its energy into the atmosphere without producing damage on the ground. The radius of effect for soft targets is much greater than for hard targets because much less energy is required to produce damage to relatively fragile structures.

We now turn to each of the four types of targets and discuss what weapons would be most suitable for destroying them or, in military parlance, "holding them at risk." Particular attention will be given to changes in strategic thinking made possible by improvements in the accuracy of the delivery vehicle.

Soft Point Targets

The destruction of soft point targets is taking on much greater importance as more countries acquire SCUD missiles and other mobile missile systems that can carry weapons of mass destruction. The launchers for these missiles are typically very fragile so that only a small explosion is required to render them inoperable, but they are difficult to find, and once located, they can move faster than aircraft can be directed to their location. Also, any attack on a missile containing a chemical, biological, or nuclear warhead poses the danger that the warhead could explode, releasing its contents over an area that might contain population centers or that might be later occupied by friendly forces. This is a case where *more* explosive force may create a *bigger problem* than existed previously. Maybe the missile would have been used, maybe not, but an attack with a sizable explosive force would certainly cause the release of deadly material or even the detonation of a nuclear warhead.

The accuracy of missiles during most of the Cold War was so poor that a large explosion had to be used to assure the destruction of even a soft point target. If you could be sure only that a warhead would land within a mile of its target, then the amount of conventional explosives required to destroy it would have been much greater than the payload capacity of even the largest missile. While a small nuclear yield might have sufficed, the complexity of maintaining many different types of weapons in the stockpile led to the use of standard high-yield designs, even though their immense destructive force was unnecessary.

The situation today is different in several fundamental ways. First, our ability to locate and track mobile targets is vastly better than it was even twenty years ago. The development of integrated command and control systems, tied directly to signals from satellites, unmanned aerial vehicles, and observers on the ground, means that weapons can be directed in real time with a high probability of target destruction. Second, the improved accuracy of weapons delivery means that we can now target not just a mobile missile launcher, but a particular *part* of that launcher, reducing the probability of an explosion that would release harmful materials. Hitting the firing controls, the launch rails, or other vital components would render the missile inoperative. Even though the warhead would remain intact, it would be useless without a missile to deliver it. Once our forces advanced to its location, it could be safely transported and destroyed.

A very small explosion—or even the impact of a high-velocity round that doesn't contain any explosives at all—can actually be more effective against a biological weapon than a low-yield nuclear weapon. The reason is that a minimum-force approach has a higher probability of containing the bio-weapon, whereas a nuclear explosion will spread it around the countryside, often without killing the agent itself. There have been a number of articles in the press that advocate the development of "mini-nukes" for attacking biological and chemical weapons, but in fact these are the *last* weapons that one would want to use against such targets since they assure the widest dissemination of the deadly material.

A potentially more serious result of using a nuclear weapon against a chemical or biological target is that it makes the user the first to cross the nuclear threshold. Since their first use against Japan, nuclear weapons have been put into a special category—they are not like other weapons. They are capable of destroying a city, a country, or even civilization itself in a matter of hours. Once the nuclear genie is out of the bottle, once a country has crossed the threshold to use nuclear weapons, other countries may infer that they are justified in using their own nuclear weapons. This could lead to the escalation of a conventional war to a global thermonuclear catastrophe. It is unlikely that using a small nuclear weapon would make a difference—nuclear is nuclear, and the political and military consequences of its use are unpredictable.

Nonnuclear weapons are thus *more* effective than nuclear weapons against chemical or biological targets. The use of existing guidance packages in precision bombs, enhanced by automatic image recognition and self-targeting in the terminal stage, enables nonnuclear munitions to be very effective in this mission. In this case, most or all of the explosives would be removed from the bomb and new electronics would be installed to enable it to hone in on the desired part of the target.

Similar technology could be used in ballistic missile warheads, including those fired from intercontinental distances. Here the problem is more complicated since the high velocity of the warhead falling from space would cause significant damage even without any explosive payload. However, warhead speed can be reduced by means of retro rockets or deployable flaps so that the force of impact would be greatly reduced.

The biggest problem in attacking soft point targets is accurate intelligence. If we know where the target is, we can destroy it. If we don't know where it is or, worse, don't know that it exists, then even a high-yield nuclear weapon will not assure its destruction.

Soft Area Targets

Some targets are very large—airfields, naval bases, and troop deployments can cover hundreds of square miles. Nuclear weapons were assigned to these targets for two reasons. First, the quantity of conventional high explosives required to destroy them was so great that hundreds of bombers would have been needed for its delivery. Precision conventional weapons could put a runway out of action for a few hours, but they could not be delivered in sufficient numbers to eliminate the target as a future threat.

Second, air defenses would have made it difficult or impossible to get the required number of bombers over the target. Only nuclear explosives could be delivered in suf-

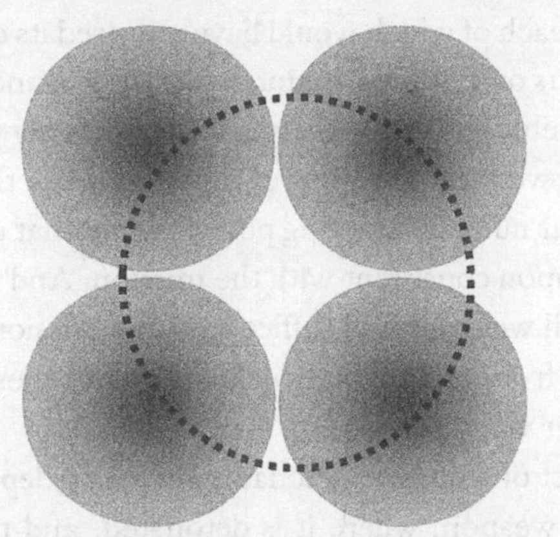

Comparison of the destructive radius of one large explosion (dashed line) versus four smaller ones, each of which has one-quarter of the yield of the large one. The smaller explosions cause equivalent damage over a larger area than the single large explosion.

ficient numbers and certainty to assure the destruction of large-area targets.

However, the most efficient way to destroy large targets is *not* via a single big explosion, but by detonating several much smaller ones distributed over the area. This is illustrated in the above figure, which shows the area damaged by several small weapons versus that of one large weapon—less total yield is required for distributed blasts than for one large one. To better understand this, think of how you might illuminate a very large auditorium—it is much more efficient to use many small lights scattered across the room than one very powerful one in the center. During the Cold War, high-yield weapons were planned for use against large soft targets not only because we didn't have enough missiles to carry many small warheads, but because we didn't want to maintain too many types of warheads—each of which would have required its own store of spare parts, its own trained maintenance crews, and its own set of delivery vehicles. Today, thanks to strategic arms reduction agreements, we have a surplus of missiles, more than enough to change our nuclear targeting policy to one that relies on the smallest weapon consistent with the mission. And we are finding that small weapons will suffice for most missions, allowing us to switch from mainly high-yield attacks to those involving much smaller weapons.

The effect of a nuclear detonation in a city depends on the yield of the weapon, where it is detonated, and the types of structures around it. The greatest damage occurs when the explosion occurs several thousand feet up—in that case the blast is spread over the widest area. Some of the force of an explosion at ground level is absorbed in nearby buildings that shadow those farther away from the full impact of the blast. However, channeling of the pressure pulse by urban canyons can complicate any predication of weapon effects. For a city consisting of nominally constructed many-storied buildings, a ten-kiloton explosion would produce severe damage over a quarter-mile radius. A one-hundred-kiloton explosion would cause the same level of damage over more than a half-mile radius, and a megaton blast would reach out to a mile and a half. Based on experience from the attacks on Hiroshima and Nagasaki, fifty thousand to one hundred thousand people would die from a ten-kiloton explosion, and correspondingly more for higher yields.

Hard Point Targets

Hard point targets include missile silos and hardened bunkers that are impervious to anything but a massive nearby explosion. During the second Iraq War, United States forces found that Iraq had built facilities underneath palaces and other civilian structures, "buildings within buildings," with roofs and walls many feet thick. Sometimes the structure was designed with multiple layers of concrete, interspersed with air gaps, to cause bombs or warheads to explode well away from the sensitive part of the facility and to diffuse their energy into the surrounding decoy structure. While aerial observers would see destruction, the actual facility would continue to operate.

Nuclear weapons were assigned to this type of target during the Cold War because it was impossible to deliver a sufficient conventional explosive force to assure destruction. To avoid having to maintain many types of warheads, it was typical to assign a much higher-yield weapon to the target than was actually required to destroy it. Another factor encouraging high yield was uncertainty in the construction of the target; an enemy seldom cooperates in providing blueprints of its facilities, so overkill by a factor of two or more was often used to increase confidence that the target would be eliminated.

Advances in weapons technology have made the use of massive explosive force unnecessary in many but not all cases. For example, the capability to place a warhead on the cover of a missile silo is no longer beyond our reach. While the amount of conventional explosive that such a warhead could carry is still less than that needed to destroy the silo, the nuclear yield required is only a few percent of that which was formerly thought to be necessary. In fact, a ten-kiloton warhead is more than adequate to cause fatal damage to the silo if placed with sufficient precision.

Many missile fields contain a number of silos connected to a central command and control facility. To destroy all of them would require one low-yield nuclear weapon per silo. One might think that accurate intelligence and guidance might enable us to destroy the command center that launches the missiles rather than each individual missile, but it is still possible that the missiles could be fired remotely, or from the silo itself. High-yield nuclear weapons may not be able to do much better than low-yield ones, since in some cases the silos were deliberately placed far enough apart that a single high-yield explosion would not destroy more than a few, leaving the rest operational and ready for a retaliatory strike. If nuclear weapons are required, then lower yields will do as well as higher ones—it is a matter of the number of weapons, not yield.

Super-hard Targets

Several facilities around the world are so hard that even a high-yield nuclear explosion may not destroy them. For example, the United States announced that it suspected Libya of constructing a WMD facility under a mountain with hundreds of feet of rock and dirt protecting its critical components. The technique used was simple: Readily available tunneling machines were used to bore a horizontal shaft into the mountain, and once they were far enough inside, the tunnel was expanded to make rooms, some of which were large enough to handle manufacturing equipment, military command posts, and weapons storage. The required mining technology for these operations is relatively inexpensive thanks to advances in civil engineering and highway construction.

Even a very large nuclear explosion—say one in the megaton range—would still not destroy a deeply buried target if its builders took certain precautions. The tunnel could be lined with hardened concrete to prevent cave-ins, and sensitive equipment could be mounted on springs to absorb shocks. This technique was used in the Cheyenne mountain control complex in Colorado, where entire buildings were built on shock absorbers to ride out a nuclear attack. Similarly, the effect of a detonation outside the entrance would be greatly attenuated if the builder put a number of twists and turns into the tunnel, each protected by a heavy blast door. Shock waves from the nuclear explosion would lose some of their energy every time they had to "turn a corner" and penetrate a blast door. Even a direct hit on the tunnel mouth would not destroy a cleverly designed facility constructed deep within the mountain.

Super-hard targets are, at first glance, impervious to even nuclear weapons. However, there are some techniques that can enhance the effectiveness of a nuclear explosion without increasing its yield. Most prominent among them are the so-called earth-penetrating weapons that bury the nuclear explosive in several feet of earth before it is detonated. Even this shallow burial increases the coupling between the output of the weapon—which is mostly in the form of x-rays—and the ground. If the weapon were detonated in the atmosphere or on the surface, then at least half of its energy would be directed upward—wasted for the purpose at hand. When the weapon is covered with enough earth to trap the x-ray energy, then much more of its energy is directed downward into the earth. The increase in efficiency could be a factor of ten or more, reducing the need for high yield or, conversely, enabling a given yield to destroy a much deeper target.

There is one remaining possibility for eliminating a super-hard target, and that is to isolate it from any contact with the outside world. No facility is built to exist in absolute isolation. Some require extensive communication connections to enable them to control military forces, while others require easy entry and egress to deploy the weapons inside them. By destroying these communication links or burying the tunnel exits, it would be possible to render the facility at least temporarily inoperative. If the intent is only to provide short-term incapacitation, then these techniques will suffice and can be achieved with conventional explosives, including those placed on ballistic missiles. However, caution must be exercised, since some exits from the facility may be hidden and there may be many redundant cable connections to the outside world. Attacks with conventional weapons may achieve some damage, but it might be impossible to determine the effect of that damage until it is too late.

IT SHOULD BE clear from this discussion that the type of weapon required for a given mission depends on the target. Small targets that are sensitive to damage—such as mobile missile launchers—can be destroyed with very little energy. Very hard targets, or those that cover large areas, require more energy. The principal determinant of success in future strategic engagements may not be the absolute power of our weapons but the accuracy of our intelligence and our ability to deliver small weapons with precision.

5

Replacing Nuclear Weapons with Advanced Conventional Weapons

Nuclear explosives are unquestionably the most powerful weapons known to humankind. However, that does not imply that they are the last stage in weapons development, any more so than were the bow, the rifle, and the airplane. Military planners are finding that explosive force is no longer the sole arbiter of military victory—a small amount of force applied at the right time and the right place can sometimes achieve the same result as a nuclear weapon. With overwhelming conventional military superiority, the United States should be the *last* country to consider the use of a nuclear weapon—once we use a nuclear weapon, we become fair game for someone else to use one against us. Nuclear weapons represent the ultimate "big stick" advocated by Theodore Roosevelt, a stick that is always understood to be present, but a stick whose primary purpose is to deter the actions of others rather than to attack.

An analogy may help to clarify the changing role of nuclear weapons in the twenty-first century. During the 1930s, the United States Navy considered the battleship the primary instrument of sea power, an essential element in checking Japanese expansionism in the Pacific. Great effort was invested in developing bigger guns and faster ships under the theory that a major naval engagement would be won by the side that could deliver a lethal bombardment outside the range of the enemy's weapons. Leviathans of forty thousand tons displacement were constructed with guns capable of shooting a one-ton projectile twenty miles or more with impressive accuracy. However, the admirals got it only half right. There was an urgent need for ships to meet the threat of Japanese expansion, but the navy needed *aircraft carriers* rather than battleships. Pearl Harbor tragically demonstrated that a single-engine aircraft carrying a torpedo or bomb could attack and sink a heavily armed ship thousands of times its size. The largest battleships ever constructed were sunk by aircraft rather than by big guns, even though a single hit from those guns would obliterate the aircraft.

JUST AS CHANGES in technology rendered the big-gun battleship obsolete, changes in technology are rendering nuclear weapons unnecessary for some military missions. There are three new technologies that can augment or even replace our reliance on high-yield nuclear weapons:

Advanced conventional weapons on ballistic missiles
Electronic and cyber warfare
Low-yield nuclear weapons on precision-delivery vehicles

Advanced Conventional Weapons on Ballistic Missiles

Some of the soft point targets that are currently targeted by nuclear weapons could be destroyed (or at least rendered ineffective) by a nonnuclear round delivered with high accuracy. For example, a SCUD launcher with a biological warhead could be destroyed by hitting the control panel, the launch rails, or other critical components of the system. Such an attack would be more effective at eliminating the SCUD threat than would a nuclear weapon, since a nuclear explosion might actually spread the chemical or biological agent over a wide area, defeating the purpose of the mission.

Some hardened underground bunkers cannot be destroyed by conventional bombs, even if they are placed right on top of the facility. But a high-velocity warhead falling from space would have sufficient kinetic energy to drive a lethal shock wave some tens of yards into the ground. The reasoning is simple: An explosion on the surface sends its energy into all directions so that only a fraction of that energy is directed downward to the underground target. Conversely, when an object strikes the surface almost all its momentum is transferred in the direction of the warhead's motion—down.

As an example, a properly designed warhead falling from high altitude can hit its target with a velocity of seven thousand feet per second or more. For a 500-pound warhead, that translates to an equivalent explosive force of 250 pounds of TNT. It may not even be necessary to put explosives in the warhead—a lump of iron would suffice, since it is the momentum that will cause the damage rather than an explosion.

There are several ways that such a warhead could be delivered. Intercontinental ballistic missiles, such as the existing

Minuteman III with a guidance upgrade, could deliver a lethal package across the globe in less than an hour from receipt of an order to launch. Land-based missiles have the advantage that the launch order could be delivered on a separate communication circuit to a separate set of missiles, eliminating even the possibility that a nuclear weapon might be launched by mistake.

A common criticism of conventionally armed ballistic missiles is that the observation of such a launch from an orbiting satellite could trigger an overwhelming nuclear response from Russia, China, or some other nation. However, one can envision communication protocols similar to those used to announce space launches, procedures that would calm the fears of the most nervous adversary. Also, it would be clear to most nuclear-capable countries that the trajectory of the single launch was not aimed at them. Finally, nuclear adversaries know that they could respond to a single nuclear-armed missile with dozens of their own. To provide even more assurances, we could have international observers at the control centers of conventionally armed missiles so that military officers from the other nuclear powers would be in continuous contact with their home countries and able to advise them of a conventional launch. These observers could even inspect the missiles to verify that they did not carry nuclear weapons.

Land-based missiles have the disadvantage that, because of the relative positions of the United States and its potential adversaries on the globe, some targets may require the missile to fly over countries that refuse permission for such overflight. For this reason we should consider putting conventionally armed missiles on submarines and surface ships, and assigning them to forward-deployed land units. Mixing nuclear and nonnuclear warheads on a Trident class ballistic missile submarine may be inadvisable due to the difficulty of assuring that there would be absolutely no chance of mistakenly launching a nuclear round when a conventional round was intended. However, the United States is modifying four ballistic missile submarines for conventional use (so-called SSGNs), and it *would* be possible to mount conventionally armed ballistic missiles on these ships with no danger of nuclear confusion. Just as reassurances can be provided to jittery countries for land-based launches, similar reassurances could be provided for submarine-based launches.

Two types of missiles could be considered for naval use—the capable and proven Trident D5 and a new or modified missile, perhaps one with a shorter range. The D5 missile was designed to be launched from Ohio class ships, and it has an outstanding record of reliability and great flexibility in targeting. It can be fired from any ocean in the world and has an accuracy that, combined with appropriate terminal guidance systems, would be adequate for conventional missions.

Short-range missiles could be fitted into the launch tubes of the converted SSGNs and could also be mounted on surface ships such as cruisers and destroyers. The Standard Missile, venerable but quite capable with its many upgrades, is one possibility, as are modifications of existing land-based missiles. One could also consider the design of a purpose-built missile for precision-strike applications. A potential limitation to new conventionally armed ballistic missiles is the proscription, signed into a treaty between the United States and Russia in 1988, against the deployment of missiles that have intermediate ranges between nineteen hundred and thirty-four hundred miles. This agreement, a relic of the Cold War, could be modified or dropped altogether since its motivation was to forestall the installation of nuclear-tipped missiles in Europe rather than conventional missiles aboard ships.

Ground units could be equipped with short-range missiles that regional commanders could control. These weapons would suffer few or none of the potential political problems associated with the launch of land- or sea-based intercontinental ballistic missiles, and with modern command and control systems, they would be an incredibly valuable tool for eliminating high-value, time-urgent targets.

Air delivery of high-velocity weapons, or other advanced conventional ordnance, is also a possibility. While a bomb falling from an aircraft, even one flying at very high altitude, does not achieve the impact velocity of a warhead falling from space, rocket-assisted warheads could be employed to increase free-fall velocity to a usable range. Also, aircraft can be equipped with other types of ordnance specifically designed for special classes of targets. The Defense Threat Reduction Agency (DTRA) has developed advanced explosives that perform with several times the efficiency of conventional munitions. These mixtures, termed "thermobarics," burn more slowly than a normal explosive, giving the molecules more time to release all their energy into the target. DTRA has also developed exotic chemical mixtures designed to neutralize chemical and biological weapons. These and other ideas could be used in warheads and bombs intended to destroy targets that would otherwise require a massive nuclear explosive, keeping the battle firmly on the conventional level while achieving the required military objectives.

Advanced conventional weapons cannot completely replace nuclear weapons—some targets are simply too hard to be destroyed by anything less than a nuclear explosion—but they can replace nuclear weapons in many missions and would provide enhanced capability without crossing the nuclear threshold. Today, when confronted with the immediate use of weapons of mass destruction against our cities or military forces, our only options are to do nothing or launch a preemptive nuclear attack. New technology offers better solutions.

Electronic and Cyber Warfare

Nearly every country in the world depends on electronics to wage war. Even low-tech battles involve radios, GPS location systems, and computers to help aim artillery. Modern armies use sophisticated computer networks to apply the right force in the right place at the right time. The command center of a modern U.S. Army unit looks like a space launch control facility—rows and rows of computers with people combing through real-time satellite images of the battlefield, data from unmanned aircraft, and reports from ground forces. Commanders can see tanks or infantry coming and move just the right amount of force to counter the threat, avoiding ambushes of our forces and saving huge amounts of money in not having to hold massive reserves for "just in case" scenarios.

Rendering these capabilities inoperative would seriously impair the capability of any fighting force, effectively rendering it blind and deaf. This is true for our high-tech forces, but it is even truer for countries ruled by dictators, where decisions are made only at the highest levels and where personal initiative on the part of frontline battlefield commanders is discouraged. The United States found in both Gulf wars that enemy units cut off from central command in Baghdad just evaporated into the desert—without specific instructions they were unwilling to take even the most obvious military action. In contrast, our forces are trained to operate independently if cut off from central command. We use technology as an aid to, rather than a replacement for, the capabilities of our officers and enlisted people in the field.

Jamming, or the purposeful broadcasting of an intense radio signal to interfere with communications, has been practiced since the dawn of radio. So too has eavesdropping on enemy signals in an attempt to anticipate its actions. Elaborate ground- and space-based systems have been developed to intercept signals from around the world; one reason that supercomputers were invented was the need to break the codes in which these signals are sent. Almost any signal that is broadcast can be intercepted, and given enough time, all but the most heavily encrypted can be deciphered.

Electronic warfare goes well beyond signal interception and jamming. Electromagnetic weapons—devices that emit a radio or microwave pulse sufficiently intense to destroy enemy electronics—have been fielded by several nations. Russia may have developed this technology to its highest point, especially in the form of microwave generators powered by compact high-explosive-driven power supplies. These devices use an explosive charge to compress a magnetic field, turning the chemical energy of the explosive into electrical energy that can be used to power a microwave transmitter. Some are as small as six inches in diameter and eighteen inches long and produce radio pulses of hundreds of kilowatts. Others are more than thirty-six inches in diameter and many feet long and could produce pulses with more power (for an instant in time) than all the electrical generators in the world put together. During the 1990s, Moscow cut off most of the funding to some of the Russian laboratories working in this area. To get money to live, they sold microwave weapon technology to other countries, most notably China, France, and the United States.

THE DISASTROUS EFFECT of unauthorized entry into computer networks has been repeatedly demonstrated in the commercial sector as hackers have planted viruses, worms, and other network attack software on the Internet. Businesses have lost tens of millions of dollars through downtime, lost revenue, and the need to replace credit cards whose numbers have been compromised. Experts in computer security maintain that the advantage remains with the attacker and that defensive measures such as firewalls and virus scans are temporary at best. Data encryption can help, at the expense of speed, but few organizations or even countries have the sophistication to create an essentially unbreakable code.

There are three approaches to computer network attack: surveillance, modification, and destruction. Surveillance is intended to find out what the adversary is doing, such as what orders are being given to whom and where key targets are located. Skillfully conducted, computer surveillance can be performed without the adversary ever knowing that it is happening. The value of such clandestine surveillance can be immense—if you know that a given military unit is being sent to a given location, then you can deploy your own forces to counter it, eliminating the necessity of searching, keeping large numbers of your own troops in reserve, and so forth.

In addition to just listening, it is also possible to inject your own information into an adversary's network, right up to the point of directing enemy troops into a trap or ordering them to stand down from attacking your own forces. The ability to mimic orders from central command requires considerably more skill than simple surveillance, since orders may be encrypted or may contain special authentication codes. However, amateur hackers have successfully inserted messages on the commercial Internet, despite security measures installed specifically to make it impossible for them to do so. Professionals can presumably do even better.

Destruction of computer networks can take the form of erasing data on storage media, disconnecting vital nodes, and inserting a virus that will propagate through and incapacitate the network. It is not necessary to physically destroy the equipment—it might be sufficient to make it inoperative during critical periods. Denying an enemy the opportunity to communicate with and control its forces can sometimes render them ineffective and accessible to attack.

The fundamental problem with electromagnetic weapons of all varieties is that they typically produce a "soft kill," or one that is not always observable from a distance. Perhaps the microwave weapon or the computer attack was successful, and perhaps it was not—how much credence would a commander be willing to give a theoretical probability when lives are at stake? Also, microwave and other electromagnetic weapons more frequently result in "upset" rather than destruction. The equipment is not destroyed, but computers need to be rebooted and communication links reset. While this type of upset could be catastrophic in a modern fighter jet moving faster than the speed of sound, it might be only an annoyance in a field artillery unit that could rapidly rejoin the fight.

Low-yield Nuclear Weapons on Precision-delivery Vehicles

In the previous chapter, we noted that very few targets require more than ten kilotons of explosive energy for their destruction. This is a small fraction of the energy of most strategic weapons, the yields of which are often in the one-hundred-kiloton to megaton range. However, fearing that any work on new nuclear weapons, even those with lower yield, could ignite a new nuclear arms race, Congress has been reluctant to fund the development of weapons that are more attuned to the strategic realities of the future. One argument used by opponents of change is that, by reducing the yield, nuclear weapons become more "usable" and present a greater temptation to military commanders to cross the nuclear threshold. (Recall that commanders can merely recommend—only the president of the United States can authorize the use of nuclear weapons.) This fear has been exacerbated by some defense analysts who argue for a continuum between conventional explosives and nuclear weapons, with the implied notion that low-yield nuclear weapons might indeed be usable if the collateral damage associated with them is made sufficiently small. I believe that these arguments are seriously flawed and fail to appreciate the essential elements of strategic deterrence. I strongly believe that the United States should do everything in its power to *increase* the nuclear threshold, to *increase* the mystique associated with a nuclear detonation, and to *increase* the fear that any country that uses a nuclear weapon outside of the direst circumstances of national survival is committing a grave international crime. This is more than altruistic thinking—it is in our practical interest to limit the use of the only weapon that could inflict tactical defeat on our forces. The United States can win any battle, any time, so long as the adversary does not employ nuclear weapons against us. One way to make sure that it does not is to tirelessly emphasize the fundamental difference between nuclear and conventional weapons and to stress that any use of nuclear weapons could trigger an immediate and devastating response on the part of the United States.

One notion that gains occasional notoriety is the idea of a "micro-nuke," or a very small nuclear explosive whose yield is measured in tons or tens of tons. This type of weapon falls into the category of minimal utility and maximal cost. A nuclear explosion of this yield does not produce effects substantially greater than those that can be achieved with advanced conventional weapons, but it gives the user the stigma of having crossed the nuclear threshold, thereby legitimizing the use of much larger nuclear weapons against it. It is unfortunate that such proposals continue to cloud the real debate on the future of our nuclear forces.

Having said this, for deterrence to have any value there must be a perception, on both sides, that nuclear weapons *will be used* in certain extreme circumstances and that they will function as designed if called upon to do so. Those with a legal turn of mind sometimes attempt to pin down governments on the exact conditions that would prompt nuclear use, failing to realize that another part of deterrence is a purposeful ambiguity that keeps an adversary from taking risks. Suppose another country mounted a biological attack on the United States. Would we respond with nuclear weapons? Perhaps, perhaps not, but the attacker would not know beforehand and might shy away from the idea of an attack just on the *possibility* that we might use nuclear weapons. It may be that we can handle the problem through conventional means, by destroying their weapons facilities so that they cannot mount a second attack. Using a nuclear weapon in that case would be unnecessary. On the other hand, if we had unambiguously said that any biological attack would trigger a nuclear response, the enemy would put us into the position of having to carry out that threat or lose face, decreasing confidence that we would ever use nuclear weapons and emboldening them to try more aggressive attacks in the future.

It is ironic that those who most object to the existence of

nuclear weapons steadfastly insist that our unnecessarily destructive weapons remain in the stockpile and that lower-yield systems should not be deployed. Congress has repeatedly cut off funding for research and development of weapons with lower yields or those with added safety features. Anti-nuclear groups have sought to characterize any change in our nuclear stockpile as tantamount to a unilateral resumption of the nuclear arms race, forgetting that Russia has announced that it is actively pursuing new weapons technology and deploying new nuclear weapons on new missiles. It seems to me that these arguments are forcing the United States to continue a policy of mutually assured destruction, one that uses much greater force, with potentially far greater civilian casualties, than is necessary.

Intelligence

Perhaps the single most important element of the future strategic equation is accurate and timely intelligence—no amount of precision or explosive power will make up for not knowing the existence and location of dangerous targets.

The United States spends approximately $40 billion per year on intelligence, but we have failed to predict many of the history-changing events of the past three decades. No major intelligence agency confidently predicted the demise of the Soviet Union, the terrorist attacks of September 11, 2001, or the correct status of weapons of mass destruction programs in Iraq. Much has been written about the causes of these failures, but most of the problem seems to rest in three areas: the value attached to information obtained via clandestine means, the lack of scholarly analysis, and the increasing technological capability of other countries.

First, many intelligence agencies still believe that their job is to uncover *secrets* when their most important mission should be to gather accurate *information*. Analysts still associate the value of a piece of information with the means that was used to obtain it—a photograph obtained by a secret spy satellite is often considered more important than a picture clipped from a foreign newspaper. But even the most sophisticated spy satellites can be spoofed, as the Serbs did during the Balkan wars by constructing plywood silhouettes of tanks and troop carriers.

Sometimes governments publish information that they want other people to know, as when the Indian BJP party announced that, if elected, it would perform a nuclear weapons test. Many American analysts rejected this information because it had not been "confirmed" by intelligence sources, missing the point that the government was making an important statement right out where *everyone* could read it. Not everything in the newspaper is true; neither is everything collected by secret means. The goal of intelligence is to sift through *all* the information available, seeking out consistencies and obvious fallacies to construct the best picture possible.

We must also be more sensitive to the *type* of information that we seek. There was a time when it was imperative for the United States to know how many missiles the Soviet Union had, their location, and their operational status. The Soviets refused to provide this information, so we made huge investments in satellite systems so that we could get it ourselves by what became known as "national technical means." Today, many of our most serious security issues involve the *intentions* of people, information that is not observable from space with even the most sophisticated satellite. What motivates a young man or woman to become a suicide bomber? How can we anticipate the attack of insurgents? One modern management fad insists that "if it can't be measured it doesn't exist," a manifestly ridiculous notion when applied to the modern security environment, where it is often the will of a few individuals, rather than physical numbers or weaponry, that is most critical in determining their actions.

The second gap in modern intelligence relates to inadequate analysis of the data that we do have. During the 1990s, the intelligence community lost some of its best analysts—many of whom had spent decades studying particular parts of the world—to retirement, resignation, or reassignment. It is unfortunate that modern methods of personnel management sometimes shift people within an organization faster than they can develop expertise in any one area, leading to poorly informed judgments that could have disastrous consequences. The fundamental failure to project the status of Iraqi WMD capabilities was partly due to this practice, but one can easily imagine the opposite danger of not recognizing a real problem in its early stages. I have seen intelligence reports that missed key points because the people who wrote them lacked the technical knowledge to understand what they were analyzing. Conversely, I have read reports that made the adversary ten feet tall—essentially superhuman—by assuming that he had much more capability than he actually had. The ability to critically evaluate intelligence data takes time to develop, and the intelligence community must recognize the need to invest over the long term with individuals who can weigh the pros and cons of each argument, people who are able to distinguish between what is *known* and what is *inferred*.

Finally, modern intelligence requires a higher level of technical competence than ever before. Gone are the days when we could hire a retired air force test pilot to get an informed opinion of the latest Soviet fighter jet. Today we are assessing a country's ability to do gene splicing, nanotechnology, and other high-tech activities. While many agencies bring in nationally recognized experts to help in technical evaluations, we must strengthen the ties between the intelligence community and the academic and industrial research sectors to provide the maximum assurance against technological surprise.

This is an urgent call to action—if we know that a danger exists, we have the capability to eliminate it. However, if we do not know that it exists, then even our most powerful weapons are of little use. As we found in Iraq, misinterpreting intelligence can have disastrous consequences. It is the policy of the United States to use force only as a last resort, and then to use the minimum force required to accomplish the mission. Better intelligence will advise us on the nature of the threat, and advanced weapons technologies will enable us to neutralize that threat with minimal damage to both sides.

6

Nuclear Proliferation

On October 13, 1960, John F. Kennedy warned, "There are indications because of new inventions, that ten, fifteen, or twenty nations will have a nuclear capacity, including Red China, by the end of the Presidential office in 1964. This is extremely serious. . . . I think the fate not only of our own civilization, but I think the fate of the world and the future of the human race, is involved in preventing a nuclear war." He said this at a time when only a few countries had the financial and technical resources to create their own nuclear arsenals, but he knew that the spread of peaceful nuclear technology meant it was only a matter of time before the nuclear club grew to alarming size. And it wasn't just opportunity—motive also played a role in proliferation. With tensions between the superpowers at a fever pitch, how certain could any country be that *it* would not turn into the next battleground between capitalism and communism? Having nuclear weapons of one's own would be a powerful motivation for the superpowers to find another place to compete.

Fortunately, Kennedy's pessimistic projection did not come to pass. Only a few countries have chosen to develop nuclear weapons, a result of concerted efforts between the nuclear states and the United Nations to create powerful disincentives for would-be nuclear weapons builders.

During and immediately following the Second World War, American and Soviet nuclear technology was surrounded by security barriers intended to keep its secrets from as many as possible for as long as possible. The hope was that only a few technologically advanced countries could duplicate the feats of the superpowers and that every effort should be made to keep the Pandora's box of nuclear know-how firmly shut. By the early 1950s, President Eisenhower realized that this policy would ultimately fail. For one thing, the United States and the Soviet Union did not have a lock on scientific talent—as the war-ravaged world rebuilt its industrial plant, it would also rebuild its research and development capability, including areas touching on nuclear energy. Eisenhower sensed the futility of trying to keep nuclear technology a secret in perpetuity and decided on a radically different course, one that traded the ephemeral assurance of secrecy for the transparency of international cooperation. In a 1953 address at the United Nations, he announced his "Atoms for Peace" program, which promised nuclear technology to any nation that promised to forswear the development of nuclear weapons. He thought it better to have countries inside the tent where they could be observed than to have them outside working on their own secret projects. What became the International Atomic Energy Agency (IAEA) was established to enable worldwide monitoring of nuclear facilities.

The United States State Department was working behind the scenes to convince countries that it was in their best interests not to develop nuclear weapons. American military planners created the notion of a "nuclear umbrella" for Europe, Japan, Taiwan, and other countries to assure them that, in the event of attack, all means would be used for their defense, including the use of nuclear weapons. In Europe, nuclear weapons were stationed on the border separating East and West Germany, creating a "use them or lose them" mentality that made the development of indigenous weapons unnecessary. It is a tribute to American diplomacy that so many countries that might otherwise have gone nuclear were convinced to remain under the nuclear umbrella of the United States.

Such assurances have not always been enough. In the past decades we have seen the nuclear club expand its membership by the addition of India, Pakistan, North Korea, and, it is widely believed, Israel. Each of these countries chose to deploy weapons for reasons associated with its own national security. India and China were at war several times in the twentieth century and continue to have border disputes. Pakistan, a Muslim state partitioned out of mainly Hindu India, fears an escalation of decades-long border disputes involving Kashmir. North Korea has sought respect on the world stage and has used its nuclear program as a bargaining chip for aid and assurances that it will not be attacked. Israel is surrounded by Arab states bent on its elimination, countries with chemical warheads and ballistic missiles able to reach Israeli cities. Some countries, particularly those whose borders are in dispute, view nuclear weapons as symbols of national legitimacy, a guarantee against the loss of territory in disputes whose origins date back to colonial rule or before.

Critics of American nuclear policy claim that continued support for a large nuclear arsenal is an incentive for other countries to construct their own weapons, but history demonstrates that countries act in their own national interest, with only secondary consideration given to what America does or does not do. None of the nuclear proliferants said that they developed nuclear weapons because America maintained a much larger arsenal or because we had been slow in carrying out arms reduction agreements.

This is not to say that American actions have no influence in the international community. For example, if the United States were to publicly announce that we were embarking on a massive expansion of our nuclear forces, then we could count on swift and commensurate reactions from Russia, China, and perhaps other countries. However, if the United States were to announce just the opposite—that we were going to *dismantle* all our nuclear weapons and allow other countries free access to all our facilities to see for themselves—I doubt that all other countries would follow suit. Even if every country were to promise to eliminate all its nuclear weapons, the wide availability of nuclear technology means that we could never be sure that somewhere, some country was not keeping one or a few weapons in reserve, a capability that could shift the strategic balance in the event of conflict. Nuclear weapons cannot be uninvented. Their permanence challenges us to handle them in a way that maximizes their deterrent value while minimizing the probability that they will ever be used. If this sounds like a contradiction, it only recognizes that international affairs seldom follow a clean, logical path in which choices are unambiguously clear.

PROLIFERATION HAS NOT stopped with India, Pakistan, and North Korea. Iran continues to defy international pressure to open its nuclear program to comprehensive inspections and seems to be pursuing uranium enrichment to much higher levels than would be required for a civilian power program. After the North Korean nuclear test of 2006, some Japanese wondered if their country, the only one to have experienced nuclear attacks, should have its own arsenal.

It is encouraging that some countries that started down the nuclear path chose to reverse course. Sweden, Switzerland, Argentina, and Brazil all flirted with nuclear programs, and all decided to abandon them. South Africa, which actually produced atomic bombs, elected to dismantle its weapons and the industrial plant that produced them. The reasons for these reversals are complex, ranging from the economic impact of an expensive nuclear weapons program to a desire to be seen as a responsible country willing to live in peace with its neighbors. Nuclear proliferation is not unidirectional—given the right conditions and incentives, it is possible for a nation to give up its nuclear aspirations. Lecturing—particularly by countries such as the United States and Russia who have no intention of giving up their own nuclear arsenals anytime soon—will certainly not convince a determined proliferant. A country must understand that it is safer and more secure without nuclear weapons than with them. This argument must involve as many nations as possible so as to lessen the fear that a compliant state would be caught unawares—and at a severe disadvantage—should one of its neighbors decide to renege on its promises. The United Nations can and must take a leading role in these discussions.

THE NUCLEAR NONPROLIFERATION TREATY (NPT) of 1970 tried to give just this type of worldwide assurance. It was designed to halt the spread of nuclear weapons while making the peaceful benefits of nuclear technology available to all nations—the same goal as the Atoms for Peace program that President Eisenhower instituted almost two decades before. Article V of the NPT obligates nuclear states to assist other countries by providing nuclear reactors for energy, for the production of special isotopes for medical diagnostics and treatment, and for industrial uses. In return, signatories agree not to use that technology for weapons development.

An inspection regime is included in the NPT to allow verification visits to nuclear facilities, including continuous closed-circuit television monitoring of key areas to ensure that they are not being used for military purposes. It is led by the International Atomic Energy Agency and involves experts from many

Article V

Each party to the Treaty undertakes to take appropriate measures to ensure that, in accordance with this Treaty, under appropriate international observation and through appropriate international procedures, potential benefits from any peaceful applications of nuclear explosions will be made available to non-nuclear-weapon States Party to the Treaty on a nondiscriminatory basis and that the charge to such Parties for the explosive devices used will be as low as possible and exclude any charge for research and development. Non-nuclear-weapon States Party to the Treaty shall be able to obtain such benefits, pursuant to a special international agreement or agreements, through an appropriate international body with adequate representation of non-nuclear-weapon States. Negotiations on this subject shall commence as soon as possible after the Treaty enters into force. Non-nuclear-weapon States Party to the Treaty so desiring may also obtain such benefits pursuant to bilateral agreements.

Article VI

Each of the Parties to the Treaty undertakes to pursue negotiations in good faith on effective measures relating to cessation of the nuclear arms race at an early date and to nuclear disarmament, and on a Treaty on general and complete disarmament under strict and effective international control.

Articles V and VI of the Nonproliferation Treaty. Article V promises the benefits of peaceful nuclear technology to any country that forswears the development of nuclear weapons. Article VI pledges the nuclear powers to work toward the elimination of all nuclear weapons.

nations to help reduce the perception that bigger countries are simply imposing their will on the nuclear have-nots.

Inspections can calm many fears, but they are not the only answer. Russia built a major nuclear power plant in Iran that, while completely within the bounds of the treaty, raises concerns about whether its real purpose was to train Iranians in nuclear technology. Critics charge that there is little reason for Iran, which has some of the richest oil and gas deposits in the world, to spend scarce national treasure on nuclear power plants. They note that Iran wastes more energy by burning off gas from oil wells than this nuclear power plant will produce.

Article VI of the NPT commits countries to work toward the eventual dismantling of their nuclear arsenals and constitutes an agreed-upon path to a nuclear weapons–free world. Every five years there is a conference to discuss the status of the NPT, attended by all the signatories. Successive conferences have seen increasingly vocal calls for the nuclear weapons states to honor this aspect of the treaty. To date, none of the "permanent five"—the United States, Russia, Great Britain, China, and France—have provided a timeline for the elimination of their nuclear forces, and it is difficult to imagine them doing so anytime soon. All have plans to maintain and modernize their weapons systems at least through 2040.

Nuclear technology has spread from its secretive beginnings to a point where most major universities in the United States and those in many other countries have a nuclear reactor for research purposes. Most of these reactors are small and contain limited amounts of low-enriched uranium (not suitable for nuclear weapons), but in combination they represent a massive amount of nuclear materials that must be protected and monitored. More serious threats for proliferation are nuclear power reactors, which contain much greater quantities of uranium and plutonium. Materials from these reactors are regularly shipped around the world for reprocessing and disposal, a growing security concern given the potential for hijacking and piracy. What if a nation-state, unwilling to spend the time and money to develop its own nuclear materials production plant, were to try a shortcut by diverting materials from peaceful uses, perhaps by the use of clever accounting or perhaps by simply withdrawing from the NPT, as was done by North Korea? Because nonproliferation consists of *not* doing something—not turning a peaceful nuclear program into one intended to produce weapons—all responsible countries must be constantly vigilant to ensure its success.

Nuclear experts who appear on news programs are fond of pointing out that both of the fundamental concepts of nuclear weapons design—the gun-assembly method and the implosion method—can be found in any encyclopedia and that more still can be found on the Internet. What these experts often fail to mention is that, as with most sophisticated technologies, the devil is in the details. Thankfully, it is not easy to make a nuclear explosive, and the checkered experience of all nuclear nations stands as testimony to the difficulty of the enterprise.

There are three distinct challenges associated with constructing a nuclear weapon. The first is to gather the required amount of highly enriched fissionable material. Uranium 235 and plutonium 239 are the common choices, although other, more exotic, heavy elements might be used. (Considerable difficulties accompany the use of some of these nonstandard materials—there are reasons why they are not used by the existing nuclear powers.) Uranium is a naturally occurring element found around the world, but most of the metal found in mined ore consists of the isotope U238 rather than weapons-usable U235. To obtain the latter it is necessary to separate the

Downward view into the Pulstar nuclear reactor at North Carolina State University. Reactors similar to this are used for studies in material science, medicine, and other fields of research.

two isotopes, a challenging task since the two have identical chemical properties, preventing them from being separated by simple industrial processes. The most common way to achieve separation is to convert the uranium metal into a gas, typically uranium hexafluoride, and inject it into a series of ultrahigh-speed centrifuges. The heavier U238 will concentrate around the outer edge of the centrifuge, enabling the lighter U235 to be extracted from the interior and processed into metal. Many hundreds of centrifuges are needed to get appreciable quantities of weapons-grade material, and a separation facility that

A bank of centrifuges used to enrich uranium at the gas centrifuge enrichment plant in Piketon, Ohio. Each stage is relatively inefficient, so many hundreds are required to separate enough uranium for a single nuclear weapon.

will produce several weapons' worth of enriched uranium per year is typically warehouse-sized. Many countries understand the basic principles of centrifuge operation, but relatively few have the technological sophistication to actually make them. High-strength steel, machined to the most exacting standards, must be used to keep the centrifuge rotors perfectly balanced as they rotate at many thousands of rpm. Reliable electrical supplies are required to avoid even momentary shutdowns of the complex machinery.

Plutonium presents its own set of challenges, principal among them that it does not exist in nature—it needs to be created atom by atom in a nuclear reactor. Uranium rods are inserted into the reactor, and radiation converts a small fraction of the uranium into plutonium. When enough plutonium has been accumulated, the rods are removed, and after a cooling-off period, dissolved so that chemical processes can separate the plutonium. Uranium can be handled with standard industrial safety precautions, but plutonium separation involves the disposition of highly radioactive waste that could kill or disable workers if adequate precautions are not taken.

Once they are separated to the required purity, uranium and plutonium must be machined to tight tolerances to ensure a fit with the other components of the weapon. Each of these materials presents its own set of unique engineering problems. Uranium is one of the hardest materials known—it is used in armor-piercing ammunition—and is exceptionally difficult to machine. Plutonium is radioactive, both on its own and because of inevitable impurities that accompany its creation in

A glove box for the handling of plutonium. Workers remain outside the box and perform work by means of the heavy gloves seen in the photo. Plutonium metal is both poisonous and radioactive and will react to air after prolonged contact.

a nuclear reactor, and it must be handled with special care lest the machinists die from exposure. Even those willing to suffer martyrdom face daunting challenges in handling this unique metal. Plutonium can exist in several different solid phases whose densities vary enough to cause significant deviations in weapon performance. Extraordinary care must be taken at each step of the fabrication process to ensure that the metal does not jump from one phase to the other during processing, ruining the component. One must also be careful to avoid corrosion. Left exposed to air, plutonium will oxidize before your eyes, and other chemicals can cause it to crumble into dust. All the nuclear weapons nations have solved these problems, but only through expensive and time-consuming trial and error.

The second challenge in nuclear weapons development is the construction of a workable "device," the euphemism given to a prototype nuclear explosive intended to verify the basic operation of the design. While some nuclear weapons experts claim that making a primitive nuclear explosive is as simple as shooting two slugs of uranium against each other in an old artillery barrel, in reality it is much more complicated. Device performance is very sensitive to the tolerances of the parts—if the slugs fit too tightly in the barrel, they are likely to jam upon being shocked by the explosives. If the gaps are too big, there is the possibility that explosive gases will get around the slugs, creating a cushion between them that will prevent any chain reaction from occurring. The yield of the explosion will depend on how fast the slugs come together, the thickness of the barrel used, and many other details of the design. It is entirely possible to design a simple gun-assembled nuclear device that fails to work. While it is true that the atomic bomb dropped on Hiroshima was a gun-assembled design that had not been previously tested, that device was the product of some of the greatest physics and engineering minds in the world, and its success should not be used as a demonstration of the ease with which a nuclear weapon might be created.

IMPLOSION DESIGNS ARE much more complicated than gun-assembled weapons. They require that the high explosive surrounding the plutonium be detonated nearly simultaneously around its outer surface to create the pressure pulse needed to compress the nuclear core. Most commercial detonators are not made with anywhere near the required precision, so special technologies need to be developed and tested with accurate measurement equipment. Also, to obtain a symmetric implosion the explosive has to be made to a level of purity much higher than is needed for commercial or military applications. Finally, the issue of gaps and tolerances must be considered. Different materials expand and contract differently with temperature. If you simply press them together, it is likely that one of them will break—much as freezing water will crack a copper pipe. Even advanced nuclear states have designed implosion devices that have failed their first tests.

The third major step in constructing a nuclear weapon is to make the "device" sufficiently rugged to endure the rigors of being dropped from an airplane or launched on a missile. It is one thing to make a handcrafted masterpiece; it is quite another to make that masterpiece strong enough to sustain wide swings in temperature, as one finds in a bomb carried under the wing of an airplane, or the strong vibrations associated with missile flight. The simple issue of mounting brackets on an implosion design can require extensive testing, since if they are too stout they can impair the performance of the weapon, and if they are too weak they could cause the nuclear explosive package to break loose or be damaged during flight.

All these technologies have been mastered by the nuclear states, and the inexorable spread of science and technology around the world means that more and more countries will be able to duplicate them in the future. However, nuclear weapons development still requires the resources of a nation-state. To think that a terrorist group, working in isolation with an unreliable supply of electricity and little access to tools and supplies, could accomplish such a feat is far-fetched at best. Even if they decided to dispense with the need for engineering the device for air delivery in favor of detonating their handmade weapon on the back of a truck, they would still need to overcome the daunting problems associated with material purity, machining, and a host of other issues that are not described in any encyclopedia or on any Internet Web site.

THE FACT THAT nuclear weapons are *not* easy to make is demonstrated by the setbacks that all the nuclear weapons states have experienced in their well-funded (typically several billion dollars per year) and nationally supported programs. As mentioned in chapter 2, French scientists had a difficult time creating a nuclear explosion despite the fact that they knew the general principles of weapons operation. Their early weapons were bigger and heavier than contemporary Russian or American designs because they could not figure out how to solve the myriad technical problems that distinguish a sketch from a blueprint. It was only after an extensive series of nuclear tests, conducted first in Africa and later at their Pacific test range near Tahiti, that they reached a level of sophistication comparable to that of the other nuclear powers.

India and Pakistan provide two more examples of the challenge of turning ideas into reality. While India conducted a "peaceful" nuclear explosion in 1974, the device was massive, much too large to fit on a rocket or small aircraft. Twenty-four years later, India conducted a series of tests, including what it claimed to be a successful hydrogen bomb. No photographs of the explosives were shown, and internationally measured yields of the tests were not consistent with what the Indian government announced. There is some doubt as to what was actually achieved.

Pakistan took a different route, reportedly receiving significant assistance from outside in the development of what is a small but apparently functional nuclear explosive. Little is known about the Pakistani device, but it is clear that the country has the capacity to produce a nuclear explosion and to measure at least some of its performance parameters.

While proliferation by India and Pakistan was unfortunate, far greater damage may have been done to international security in the 1980s by A. Q. Kahn—the so-called father of the Pakistani bomb—and his wholesale marketing of nuclear weapons technology. There are reasons to believe that Libya, Iran, and North Korea were Kahn's eager customers, and it is likely that terrorist groups, including al Qaeda, expressed interest as well. The ultimate value of this information to the recipients will depend on their ability to turn sketches into hardware and to assemble and operate the complex equipment that is purchased as part of Kahn's do-it-yourself kits.

KEY TO THE success of all nuclear powers has been the ability to conduct one or more nuclear tests. Nuclear testing is more than just a demonstration of success—it enables scientists to understand in detail the complex processes that occur during a nuclear detonation. Nuclear weapons operate at temperatures of many millions of degrees and pressures of millions of atmospheres, conditions well beyond what can be achieved in a laboratory. Scientists have theories of how materials behave under such conditions, but they are just that—theories that are not backed up by solid empirical data. One way to impede the proliferation of nuclear weapons is to keep them from being tested, either as entry-level gun-assembled devices or as more sophisticated hydrogen bombs.

Several treaties govern the testing of nuclear weapons. Responding to growing concerns about the effects of nuclear testing on public health, many countries signed the Limited Test Ban Treaty of 1963 that prohibited nuclear explosions in the atmosphere, in space, or in the ocean; only underground tests in which most or all of the radiation was contained were allowed. The Threshold Test Ban Treaty of 1974 further constrained underground explosions to a maximum yield of 150 kilotons with provisions for confirmatory measurements by Russia on American tests and America on Russian tests. In 1992, President George H. W. Bush announced a moratorium on further nuclear tests by the United States, a temporary measure that has been honored ever since. The last U.S. nuclear detonation occurred on September 23, 1992.

The Comprehensive Test Ban Treaty (CTBT) seeks to make the cessation of nuclear testing permanent and universal. While many countries have ratified the CTBT, the United

States Senate refused to do so, citing concerns that America might not be able to fix unanticipated future problems in its nuclear stockpile or develop new weapons required to deal with changes in international relations. The Senate did not take this decision lightly, and the senators were well aware of the political implications of the United States declining to accept restrictions adopted by all the other nuclear nations. After intensive and highly technical discussions conducted at the highest security levels, they concluded that they were not yet confident that the United States could maintain its nuclear arsenal indefinitely without testing. Rather than place the country in the position of having to withdraw from a ratified treaty, they chose not to give their consent.

The CTBT illustrates a fundamental divide between groups discussing the future of nuclear weapons. Some believe that a proactive stance by America in the elimination of all nuclear weapons, or the reduction of our stockpile to as few as one hundred weapons, would provide an example for other nations to follow. They reason that the United States has little to fear from any military threat and that now is the time to reverse the course of nuclear armament. Others insist that countries pursue nuclear weapons for their own perceived needs and that disarmament by the United States might actually encourage others to develop a decisive strategic nuclear advantage, one that could be used to blackmail or even attack us. Altruism does not always guarantee success in international relations.

AN EASIER ROUTE to nuclear status is to buy, steal, or be given a weapon ready for use. Giving away a nuclear weapon may seem irresponsible, but Russia was willing to give China substantial help in the 1950s (although it stopped short of provid-

Photo of a nuclear test emplacement at the Nevada Test Site. The tower houses a cylindrical rack that contains the nuclear device and the experimental diagnostics. This assembly will be lowered into a predrilled hole many hundreds of feet deep, which will then be filled with a precise mixture of materials to prevent any leakage of radioactivity following the detonation. The cables in the foreground will carry the signals from the experiment to the surface.

ing a detailed design) and China is reported to have provided key assistance to Pakistan in the 1990s. One of the parade floats celebrating the detonation of the first Pakistani atomic bomb in May 1998 was emblazoned with the phrase "The Islamic Bomb," suggesting that some countries may not be afraid to help others gain membership in the nuclear club.

Another option for an aspiring nuclear state or group would be to steal a nuclear weapon, perhaps from one of the Russian storage sites that are reported to have poor to nonexistent security. Press photographs of broken fences guarded by hungry recruits with antiquated rifles are scary, but the truth is sometimes different from what is reported. One journalist quoted a Russian man as saying that he "jumped the fence" at the nuclear weapons laboratory at Sarov to visit his girlfriend. However, since the "fence" at the facility consists of a triple array of fences nearly twenty feet tall, regularly patrolled by well-armed guards with dogs and continuously monitored by sophisticated sensors, this individual was either lying or was an athlete of superhuman ability.

It is true that security at Russian nuclear storage sites needs improvement, but there are several programs actively working to improve the situation. The Cooperative Threat Reduction (CTR) program, a joint effort between Russian and American technical experts, has built new fences, trained and equipped guards, and consolidated nuclear weapons at fewer locations that can be better defended. The CTR program has assisted the Russian government in disposing of dozens of long-range bombers, ballistic missiles, and nuclear submarines in a completely verifiable manner. While progress has been slower than many would like, it is encouraging that *so much* has been accomplished. Compared to how much money the United States spent on defending itself against these same weapons, the investment in the CTR program is surely one of the greatest defense bargains in history.

Stealing a nuclear weapon or a significant quantity of weapons-grade material from a Russian storage site still requires one to transport it within and presumably outside the country. Anyone who has ever traveled any distance in the Russian hinterland knows that this is not an easy task and requires considerable insider assistance. Bribes could be made, but the more people in on the deal, the higher the probability that they will be discovered.

In the mid-1990s American security became concerned about mafia activity in the secret Russian nuclear weapons city of Sarov. It seems that a Lincoln Continental was spotted next to what was presumed to be the laboratory director's house. Where had the money come from for such a car at a time when residents of the city were having trouble putting food on their tables? During my next visit to Sarov I was able to verify that a Lincoln was indeed parked on the director's street. But it belonged to someone else with a name similar to the director's, someone who was indeed associated with "unofficial trading activity." In discussions with representatives of Russian security, I learned that they had made a deal with the mafia. Fixed quantities of cigarettes, vodka, and gasoline could be brought into the city and sold, but mafia members would stay away from anything dealing with nuclear weapons. If they violated the agreement, they would be killed. In Russia, this was not an idle threat.

Rumors of missing or stolen Russian weapons or nuclear material were chronic during the 1990s and persist today. Alexander Lebed, a retired general and Russian presidential hopeful, claimed in 1996 that several "suitcase bombs" had gone missing from army storage facilities, a serious concern for every country in the world since it was also rumored that these small "atomic demolition munitions" did not have all the safety interlocks and code provisions of larger bombs and warheads. However, in a country in which nearly everything was a closely held state secret, Lebed was not on the list of people with access to nuclear weapons information. Apparently he was repeating a rumor that he had heard without accurate information to back it up.

Russians bristle at assertions that they do not take nuclear weapons security seriously. A senior official in the Russian nuclear weapons program once told me, "We are the ones surrounded by new countries, including those with unstable governments. If anyone has a fear of the use of nuclear weapons, it is Russia." Improvements are certainly needed in security, and needed urgently, but exaggerated claims of poor security at Russian nuclear weapons sites could actually *encourage* terrorist groups to try to steal a nuclear weapon—the last thing that any responsible government would want.

TAKING POSSESSION OF a nuclear weapon does not imply that one has the capability to explode it. In contrast to what is shown in movies, nuclear weapons do not have a red button on their side with an LED display counting down the seconds to detonation. Most are tightly sealed packages with only a single electrical connector serving as their interface to the outside world. Looking at such a connector provides no indication of what wire does what—some send coded signals that prepare the weapon to detonate, but others might simply report details of weapon status. Dismantling the weapon (not always an easy task) would provide more insight, but here again, most subsystems are sealed in their own cases so that it is sometimes difficult even for an expert to identify what component does what. Of course, a weapon could be *completely* disassembled and then rebuilt with a new control system, but this would require extreme care, and in most cases an intimate knowledge of the weapon's design in order to avoid destroying key components.

To explode a stolen nuclear weapon, one must have an experienced person who knows how to operate it. Such people are very rare. Weapons designers know *how* a weapon works, but few know the types of signals (timing, voltage, polarity, etc.) that must be sent to detonate a weapon. Weapons maintenance personnel are trained to perform only a limited set of functions and are often ignorant of the details of the device on which they are working. Only a few people in the world have the knowledge to cause an unauthorized detonation of a nuclear weapon.

THE DIFFICULTY INVOLVED in constructing a nuclear weapon or using one that is stolen does not mean that we can relax about the danger of nuclear proliferation. Nuclear weapons are difficult to make, but some countries are willing to invest the required resources. The inexorable spread of technology has made it impossible to prevent proliferation by simply protecting a set of nuclear secrets, a fulfillment of President Eisenhower's fear. We must make every effort to assure that nuclear technology is used only for its intended peaceful purpose and to keep special nuclear materials away from those who would use them for illicit purposes. As difficult as this might seem, there are ways to accomplish this mission.

The starting point for containing the threat of proliferation is the uranium and plutonium that are the central components of any nuclear weapon. Security fears about Russian nuclear sites are sometimes overblown, but there is an urgent need to accelerate and complete needed upgrades to storage facilities. Squabbling between the U.S. and Russian bureaucracies has often slowed progress, as when security upgrades at one Russian nuclear site were delayed while Pentagon officials argued about how many flashlights and ladders could be bought with American money. Senior-level attention is needed on both sides to break such petty impasses and restore a sense of urgency to the protection of nuclear material.

The security of weapons in newly emerged nuclear states could be dramatically improved by the sharing of command and control technologies by the advanced nuclear nations. The

United States' initial refusal to recognize India as a nuclear weapons state had little meaning given that nation's demonstrated capability to create a nuclear explosion. Rather than hope that India will voluntarily disarm, it would be better to recognize the existence of a nuclear India and provide technology to make its weapons safer and more secure. The same may be said of Pakistan. Care must be taken to avoid making weapons more usable or more effective, but ways can be found to improve security without giving away advanced weapons secrets.

Finally, the United States and Russia must remember their obligations under Article VI of the NPT—the one that commits them to reduce and eventually eliminate their nuclear weapons stockpiles. A goodwill gesture consisting of reduced numbers, greater transparency, and perhaps lower-yield weapons may be a step toward convincing other countries that they need not embark on a nuclear path of their own.

7

Defense Against Nuclear Attack

Deterrence served as a form of nuclear defense during the Cold War—no one wanted to risk the threat of massive retaliation by initiating a nuclear attack. Nuclear weapons were owned by largely rational governments who understood the consequences of using them. Today we face quite a different set of threats, a mix of traditional nation-states that are deterred by the assurance of massive retaliation, rogue states that might not be so deterred, and international networks of terrorists that are willing to risk annihilation for the opportunity to inflict serious damage upon the United States. The inexorable advance of technology around the world has made it possible for almost any determined country to acquire at least a primitive nuclear capability and for terrorist groups to be given a weapon by a sympathetic sponsor.

We divide the problem of defense against nuclear attack into three parts: defense against *clandestine attack*, defense against *attack by aircraft and cruise missiles*, and defense against *attack by ballistic missiles.*

Clandestine Attack

Every year hundreds of tons of illegal drugs and other contraband are smuggled into the United States by land, sea, and air. Thousands of trucks cross our border every day from Canada and Mexico, few receiving more than a cursory inspection. The bulk of international trade occurs via seagoing containers, some eleven million of which enter the country every year. The Department of Homeland Security has installed detectors at ports, but we are still far from confident that nuclear materials could not be smuggled in by sea.

Hundreds of thousands of packages are shipped daily by air, again with only random inspection of what's inside. Anyone can, after a few moments' thought, come up with a viable scenario by which something as small as a nuclear weapon might be smuggled into the country and secretly placed in any of our major cities. The good news is that we can take several measures to lower the probability of success and indeed to discourage people from even trying.

Our best defense against any form of attack is good intelligence—knowing the intentions of an enemy in time to thwart its attack. One of the reasons that the individuals associated with the attacks of September 11, 2001, were able to remain hidden is that very few of them knew the full extent of the plot. Experience with clandestine activities suggests that the probability of a security leak grows when the number of people in the know exceeds thirty to fifty. Some communication, some financial transaction, some package will be intercepted to alert the authorities to the plot. Dividing the team into small parts, each operating independently of the others and ignorant of the total plan, can aid in keeping activities secret, but an operation of the size required to steal or accept a nuclear weapon and deliver it to a target would likely be large enough to arouse suspicion.

Unfortunately, experience tells us that we cannot rely on our intelligence agencies—which have been surprised time and again over the past decades—to give us advance warning of an impending nuclear attack. The unique destructive capability of nuclear weapons demands that we assume that we will *not* know about an attack and that we use every means at our disposal to detect a weapon smuggled into the country. Here we are fortunate, in that a number of existing and developing technologies can detect nuclear materials in buildings, on ships and aircraft, and even in moving vehicles. Long past are the days when white-coated men used clicking Geiger counters to detect the telltale radiation associated with nuclear materials. Sensitive detectors can be placed at tollbooths, under roadways, and at airports to sound the alarm should any nuclear material pass nearby. Other types of detectors search the huge cargo holds of container ships filled with steel containers stacked one upon the other with only inches of space in between. At the other end of the spectrum, there are miniaturized detectors hardly bigger than a cell phone that can be issued to local law enforcement personnel to wear on their utility belts.

All these are passive systems in that they detect the radiation given off by the uranium or plutonium in the bomb. Another approach is to use beams of neutrons or other subatomic particles to scan the interior of containers, vehicles, or even buildings to stimulate radioactive material into giving away its location. This is referred to as "active interrogation." Active detection systems are more effective than passive systems, but their use must be limited to avoid exposing people to unsafe levels of radiation.

An interesting cross between passive and active detection uses naturally occurring cosmic rays as the "active" beam, minimizing exposure to the public. Even though the number of cosmic rays reaching the surface of the earth is small, when one of them strikes a heavy nucleus like uranium or plutonium, it gives off a unique signature that stands out from the natural background. This method is not as effective as active interrogation, which uses a much more powerful beam to stimulate the nuclear material, but it may be better than purely passive detection.

There are limits to what can be achieved by any type of detector. When I was director of the Defense Threat Reduction Agency, Deputy Undersecretary of Defense Mike Wynne accosted me in the hallway of the Pentagon. "I want you to develop a detector that we can put onto a satellite to detect smuggled nuclear material anywhere in the world," he said. "I don't think that's possible," I responded, "for fundamental reasons." "Wrong answer, Younger!" he exploded. "Go back and think harder!"

I did go back and I did think harder. I enlisted the help of experts in the field and was able to show that even a very large detector in space—one that was tens of yards across that worked with perfect efficiency—could not pinpoint the location of nuclear material. The signal reaching orbit was simply too small to permit detection, and the amount of naturally occurring radiation in space was just too great to pick out such a tiny signal.

INTERCEPTING AND NEUTRALIZING a smuggled nuclear weapon involves more than just having the right detector—

the operational side of the problem is at least as complex as the detection. Foremost among these challenges is distinguishing between the radiation emitted by a nuclear explosive and that from innocent radioactive sources. Small radioactive sources are routinely used in many medical facilities. Capsules containing small amounts of radioactive material are used to log oil and gas wells, and it is not unusual to find several of these in use within a single large field. Radioactivity occurs in the most innocuous of places, including sand, concrete, and even kitty litter. (All these arise from natural radioactivity found in the earth.) Granite buildings emit radiation owing to the presence of the radioactive gas radon in the stone. Cancer patients returning from radiation therapy have set off nuclear detectors by the residual radiation left in their bodies. Some advanced detectors can distinguish between the types of radiation emitted by these innocent sources and those emanating from a nuclear explosive, but at present they are expensive and additional work is needed before they are sensitive enough and cheap enough to permit widespread deployment.

A further complication of detecting a nuclear weapon is that the enemy might use shielding to reduce the radiation signature from the weapon. For a sophisticated nuclear device stolen or bought from an advanced nuclear nation, a few hundred pounds of lead would be sufficient to reduce the radiation profile to nearly undetectable levels; for more primitive weapons, of the type that an entry-level nuclear power might construct, the amount of shielding could easily reach several tons, making the weapon difficult to move and hide.

However, heavy shielding could itself alert authorities that something was amiss—if the manifest lists baby clothes and the container is grossly imbalanced, then alarms could ring. A big black spot on an x-ray—indicative of materials like lead—

would also be a giveaway. There are ways around even this limitation, such as placing a nuclear device in the hold of a ship transporting thousands of tons of grain, oil, or other bulk materials, but again this begins to stretch the limits of credibility. It is much more difficult to find a nuclear device that is heavily shielded, but detectors are being developed by government laboratories to address just this possibility and tests have already been conducted by the coast guard.

Detection is only the first step in safely disposing of a smuggled nuclear weapon. We must also train first responders—police, firefighters, and other emergency services personnel—on what to do with a nuclear device after they find it. How do you anticipate and deal with booby traps in a vehicle that might detonate the bomb when the door is opened for inspection? What about the possibility of a timer on the bomb that could make the search time-urgent and preclude more thorough methods? Having a detector is only half of a practical system for intercepting nuclear materials—what happens after discovery is at least as challenging.

The Defense Threat Reduction Agency of the Department of Defense has tested a number of promising technologies for detecting nuclear materials in real-life situations. Practical demonstrations were done with various detector configurations, alarms, response forces, and disposal techniques. DTRA found that some detectors did not work as well as expected, whereas others worked much better and showed considerable promise for widespread use at civilian facilities. In one case, a roadside detector was able to identify nuclear material in vehicles passing at high speed, enabling law enforcement personnel to be notified to stop the suspect vehicle.

DTRA used some of these detectors to find and recover stolen radioactive material during the Iraq War of 2003. A highly radioactive source—one that emitted much more radiation than any nuclear weapon would—was stolen from an Iraqi military training facility. DTRA personnel mounted nuclear detectors in a helicopter that flew a low-altitude search pattern across the desert. They found the source in short order, but unfortunately only after one Iraqi died of radiation poisoning and several others were severely injured by keeping the source—which was warm to the touch—close to them.

The Domestic Nuclear Detection Office of the Department of Homeland Security built on the work of DTRA to construct a comprehensive plan for the defense of the country against smuggled nuclear material and weapons. The best scientists

A truck passing through an array of nuclear material detectors on the Department of Homeland Security test bed at the Nevada Test Site. Such detectors could be mounted at border crossings, tollbooths, or at the entrances to critical installations.

from universities and government laboratories have been engaged to develop new detection technologies. They have conducted a series of tests to verify that these technologies work under operational conditions. Experiments done at the Nevada Test Site scanned large tractor-trailer trucks as they passed a tollboothlike building—the figure opposite shows one of these systems in action. Thousands of smaller detectors have already been purchased by law enforcement, National Guard units, and other groups, and more capable detectors are in development.

Just as in any other type of smuggling, it is impossible to create a leakproof nuclear detection shield around the United States. Just as drugs and other contraband are smuggled into our country, so too could a clever and determined adversary succeed in bringing in a nuclear weapon. However, we are much better off than we were even a decade ago, and if government programs are even modestly successful, we will be safer still in coming years.

Delivery by Aircraft or Cruise Missile

Any country that has the capability to construct a primitive nuclear weapon would likely have aircraft that could deliver it, at least over short ranges. Aircraft are the preferred means for delivering nuclear weapons in both India and Pakistan, both of which are struggling to perfect ballistic missiles for nuclear use.

Short-range cruise missiles, which are available for purchase on the world arms market, could be adapted to carry a small nuclear warhead. Many cruise missiles are self-contained units launched from a truck chassis, so they could be fired from a ship outside U.S. territorial waters, solving the range problem while keeping the probability of detection to a minimum. They can be programmed to follow a preset course to their destination, flying very low to avoid detection, but their small payload capacity precludes the use of the heavy warhead designs typical of new entrants to the nuclear club. Add to this the possibility that the relatively slow missile could crash or be shot down en route to the target, and it becomes less attractive to a developing nuclear power.

For now, the most likely air delivery vehicle for a proliferant is the manned aircraft. However, they are challenged by their limited range—only a few advanced countries have military aircraft with the capability to reach the United States from their home base. Developing nations' air forces are typically limited to small fighter-bombers that have ranges measured in the hundreds of miles, planes that could not cross an ocean without refueling. There are few islands in the right places that could provide the secrecy needed for such an operation, and in-flight refueling is a skill possessed by only a few countries.

One option for solving the range problem would be for a country to use a civilian airliner to carry a nuclear warhead, but again the problem is in the details. If tensions were high enough to threaten war, commercial flights would come under closer scrutiny and might be canceled altogether, as has happened many times in the past when we closed our skies to the airlines of a suspect country. Any country that would place a nuclear explosive on one of its own commercial aircraft would know that the source of the nuclear explosion would be immediately discovered and that massive retaliation would swiftly follow. (Smuggled nuclear weapons on innocent aircraft would fall under the clandestine delivery scenario described above.)

The United States keeps close track of the airspace surrounding North America as part of activities at NORAD, the North American Air Defense Command, a joint venture between Canada and the United States that uses massive radars to track every air vehicle approaching the continent. All legitimate flights file a formal flight plan before takeoff, so NORAD can quickly identify any unexpected activity in time to query the aircraft about its intentions. If NORAD is not satisfied with the answers, fighters are sent to intercept and, if necessary, shoot down the suspect plane.

During the early part of the Cold War the United States had ground-based interceptor missiles to defend against air attack. These have long since been dismantled. Military aircraft from the Air National Guard and the air force are the primary means of intercepting suspicious flights, but portable surface-to-air missiles can be used to protect high-value targets in times of crisis. These missiles include the long-range Patriot, famous for its service during the first Iraq War, and several short-range, shoulder-fired missiles for the defense of individual buildings. Although the Patriot's record at intercepting SCUD missiles was less than perfect, it is quite effective against slower-moving aircraft.

Despite the sophistication of their radar and tracking computers, NORAD and its counterparts in other nations occasionally miss things, as was demonstrated in 1987 when a young German man flew a light plane from Helsinki to Moscow's Red Square. The Soviet Union had layer upon layer of air defenses but they were still unable to detect, let alone shoot down, the small private plane that invaded its airspace. The reason was that radar has difficulty finding aircraft flying very low (literally "below the radar") and those that hug hills and valleys to avoid detection.

Smugglers fly tons of illegal drugs into the United States every year using private aircraft big enough to carry a small nuclear explosive. Small planes suffer the same range problems as those discussed above so they would need to come into the

United States from either Mexico or Canada, again requiring a sophisticated operation in which the weapon was transported to the takeoff point and then flown the final distance.

Defense Against Ballistic Missiles

Ballistic missiles are the vehicles of choice for the delivery of nuclear weapons across great distances. They can reach their target within an hour of launch, are very difficult to intercept, and have payload capacities large enough to carry a small nuclear weapon. All the major nuclear powers have most of their nuclear warheads mounted on ballistic missiles. As we have already discussed, mounting a nuclear weapon on a ballistic missile involves some very complex engineering, but we can expect more countries to solve this problem in the not too distant future.

There are two principal challenges associated with intercepting a ballistic missile in flight: finding it and hitting it. The United States has a system of satellites that can detect the launch of a missile and track it over at least the early part of its flight, long enough for ground-based radar systems to take over. One of the methods used is to look for a very bright light source moving up from the surface. There are many bright sources of light on the surface of the earth—for example, stadium lights, fireworks, reflections of sunlight—but very few of them move with the characteristics of a ballistic missile, so discrimination is relatively straightforward. Also, most countries announce peaceful launches of research rockets and larger missiles intended to place satellites in orbit, if for no other reason than to warn aircraft to stay out of the launch area. Any unexpected launch of a missile raises immediate suspicions.

Even an announced launch can sound the alarm, as hap-

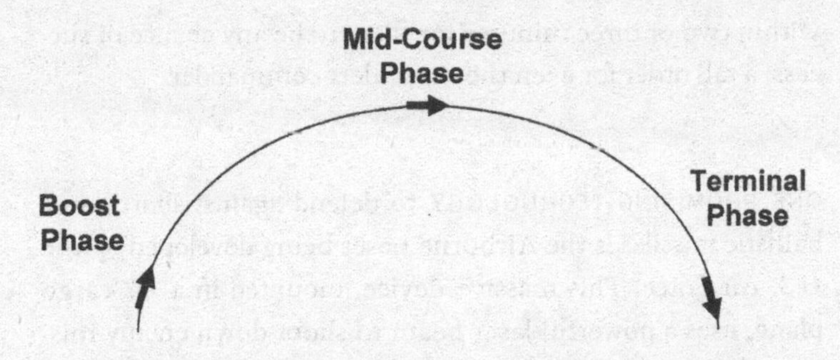

Three phases of ballistic missile flight. The boost phase is the initial ascent of the missile from the surface. During the mid-course phase, the missile is flying through space. It is during this phase that it deploys its warhead. During the terminal phase, the warhead falls to the target with velocities of about one mile per second.

pened in 1987 when Norway launched a small research rocket, and Russian forces, missing the notification, went on high alert, suspecting attack. It was only after Russian radar personnel determined that the rocket was not headed toward them that the error was corrected.

Once detected, there are three options for intercepting a missile: during the boost phase early in the flight, during mid-course when the uppermost stage of the missile is flying in space, and during the terminal phase when the warhead is streaking down toward its target. Each has its own set of advantages and complications.

The advantage of boost-phase intercept is that the missile is moving relatively slowly and is still physically large since it has not yet jettisoned its vulnerable first and second stages. The disadvantage is that the boost phase of missile flight lasts for only a few minutes, placing enormous burdens on the defender's command and control systems. The missile would have to be detected and confirmed, and the order given to shoot it down within two or three minutes for there to be any chance of success, a tall order for even the most alert commander.

ONE PROMISING TECHNOLOGY to defend against short-range ballistic missiles is the Airborne Laser being developed by the U.S. Air Force. This massive device, mounted in a 747 cargo plane, uses a powerful laser beam to shoot down enemy missiles shortly after they are launched. If a sufficiently intense spot can be kept on target for several seconds, it can burn though the missile's thin aluminum skin and detonate the propellant or otherwise so distort the metal that the missile will break up in flight. Although development has been slower than anticipated, the basic technology of aircraft-mounted lasers appears sound.

Lasers, however, have limitations. Some lasers are unable to penetrate clouds or rain, limiting them to fair-weather battles. Even clear air turbulence can disrupt the exquisite laser pulse and hence its effect on the target. Most important, airborne lasers are only effective when they are aloft near a missile launch—and they can remain on station for only twelve to eighteen hours at best. You might solve some of these problems by mounting the laser in space, where it could shoot from above the clouds, but that would be in violation of treaties forbidding the placement of weapons in space, and it would also require solving a whole other set of technological problems related to the longevity of highly complex systems in earth orbit.

The United States has placed greatest attention on defeating ballistic missiles during the mid-course phase of flight, that is, after the missile has reached space and when an accurate course has been determined by ground-based radar. From a practical standpoint, this is about the soonest that the threat can be confirmed and permission can be obtained to fire an interceptor missile. The challenge of mid-course interception is the high velocity of the missile—by that time the target is moving at speeds of several miles per second. And since it has jettisoned its first and second stages, it is only a small target in a very big sky, reminding one of the phrase "hitting a bullet with a bullet." Even the smallest error in aim would result in a miss and a catastrophe as the warhead sped unimpeded toward its target. The engineering demands associated with mid-course interception are severe, but tests have already demonstrated our ability to detect a launch and fire a missile to successfully intercept a mock enemy warhead. Knowing that mid-flight interception is *possible* is a tremendous technical advance. It may be difficult, it may take time, and it may be expensive, but it is possible.

The National Missile Defense architecture involves myriad detectors, communication systems, and interceptors intended to protect all fifty states from long-range attacks. The Space-Based Infrared Systems, one in high orbit and one in low orbit, are intended to detect launches anywhere in the world. Advanced radar systems then take over to track the target and send data to interceptor missiles in Alaska (initially) and elsewhere.

Hitting a warhead the size of a trash can moving at several miles per second was still a daunting engineering challenge when the first interceptors were being designed, but that was not the only problem facing missile defense. To deploy the new system, the United States needed to withdraw from the Anti–Ballistic Missile Treaty of 1972, a treaty that set strict limits on where and how many defensive missiles the United States and the Soviet Union could have. Although the treaty was no longer relevant to a post–Cold War international environment, Washington's announcement that we would withdraw prompted howls of protest from Russia and China. Russia threatened to field a new type of nuclear warhead that could evade interception, and the Chinese said that they would be forced to greatly increase the size of their relatively small nuclear arsenal. Both countries worried that such a shield around the United States could render their own missiles obsolete, giving the Americans the opportunity to strike without fear of retaliation.

In fact, the U.S. missile defense system contains only a few missiles that would be no match against the hundreds of Russian or even dozens of Chinese weapons. Also, the number of interceptor missiles fielded by the United States was comparable to the number already in place in the Russian defensive system around Moscow. The fundamental worry of Russia and China was that the initial American interceptor missiles were only a harbinger of more advanced systems to come, systems that could upset the delicate balance of mutually assured destruction that was the foundation of the theory of deterrence. International tempers eventually cooled, and when the United States formally withdrew from the treaty, the story rated no more than a short article on an inside page of the newspaper.

Other ballistic missile defense schemes appear promising. The U.S. Navy has tested a very capable ship-based system that would enable short-range interceptor missiles to be located off the coast of suspected trouble spots. Ships can loiter for weeks or even months, enabling them to provide an additional layer of defense to interceptor missiles in Alaska. They don't suffer the problem of remaining aloft like the Airborne Laser, nor do they have problems with clouds or rain. We must remember, however, that all these technologies were designed to thwart an attack involving one or at most a few attacking missiles—they cannot and were never intended to protect against a massive launch of the type feared during the Cold War.

U.S. anti-ballistic missile defense system being installed in Alaska in 2004 This missile is designed to destroy a nuclear weapon launched by a rogue nation.

There have been several proposals to place small interceptors in space where they would be prepositioned to attack missiles in the mid-course phase of flight. The "Brilliant Pebbles" scheme proposed by the Lawrence Livermore National Laboratory in the 1980s envisioned putting thousands of small independent rockets in low earth orbit, each equipped with onboard sensors and computers. Extravagant claims were made that these interceptors could be mass-produced at low cost and that they would be smart enough to distinguish between an enemy ballistic missile and a peaceful manned space launch.

Impressive progress was made, but the technology of the time was inadequate to the challenge.

The last opportunity to destroy an incoming weapon—terminal phase interception—occurs when the warhead is in its final free-fall from high altitude to its target. Terminal phase interception has the advantage that the interceptor need fly only a short range to the incoming missile, but it has the major challenge of hitting an oncoming warhead moving at more than one mile per second. This would be hard enough if the warhead was flying a predictable ballistic trajectory, but it becomes Herculean when the warhead has the ability to maneuver, as the Russians have claimed. Add to this the possibility that the warhead might detonate when it detects an approaching missile, and terminal defense becomes much less attractive. At best, the defensive missile would cause the enemy nuclear detonation to occur at several tens of thousands of feet rather than at the surface.

It is worth repeating that ballistic missile defense is not new—during the Cold War the United States and the Soviet Union had anti–ballistic missile defense systems that were remarkably advanced for the time. Both systems had nuclear-tipped missiles since neither had the technology capable of hitting a high-speed missile with a conventional explosive warhead. A combination of mid-course and terminal interceptors gave two chances to kill the incoming warhead. The United States dismantled its ABM system in 1976; the Soviet system is thought to be still partly operational.

The Changing Nature of Nuclear Defense

The assurance of massive retaliation was considered sufficient to deter any but the most foolhardy regime from attacking the

United States during the Cold War. Today we face a range of adversaries from terrorist groups to rogue nation-states, some of which might be willing to risk their own destruction to inflict damage upon the world's sole superpower. If the United States augments traditional deterrence with missile defenses, it sends a signal to would-be attackers that they could suffer a devastating response without their weapons even hitting America, something that might discourage them from developing nuclear weapons in the first place. Why bother to go to the trouble and expense of a nuclear capability when its chances of success are small?

Improved relations between Russia and the United States have reduced the probability of a massive attack involving hundreds of nuclear weapons. However, the spread of missile technology around the world, including major programs in North Korea and Iran, has introduced a new class of threat. The consequences of failure to detect and destroy an incoming nuclear weapon are extraordinary and suggest that even an imperfect system might be worth the investment. The United States must have active efforts in every area of nuclear defense: strong inducements to prevent proliferation, penalties if these inducements fail, treaties to control weapon types and numbers, detection programs to find any weapon that might be smuggled into the country, and active systems to shoot down air- or missile-delivered weapons.

8

Maintaining Our Nuclear Forces

The development of nuclear weapons has always been in civilian hands. During the Manhattan Project, many of the scientists working on the bomb refused to don uniforms and report through a chain of command. Major General Leslie Groves, military head of the project, succumbed to their demands and allowed civilians to direct most of the research and development. After the war, the civilian tradition continued with President Truman's formation of the Atomic Energy Commission (AEC), a group of five experienced individuals who ensured that nuclear weapons requirements were met, funds were properly expended, and safety and security were maintained. Subsequent administrations created the Energy Research and Development Administration, the Department of Energy, and, most recently, the National Nuclear Security Administration (NNSA), a semi-autonomous organization within the Department of Energy. Almost all the hands-on work relating to nuclear weapons development, from initial design to final assembly, is done by private contractors, including the three nuclear weapons laboratories (Los Alamos and Sandia in New Mexico, and Livermore in California), the four production plants (Y12 in Tennessee, Savannah River in South Carolina, the Kansas City Plant, and Pantex in Texas), and the Nevada Test Site. NNSA oversees these activities, particularly in areas such as nuclear safety, strategic planning, and the interface with the Department of Defense. Approximately $7 billion was allocated to nuclear weapons research, development, and manufacturing in 2009.

Each year the directors of the three nuclear weapons laboratories sign letters certifying the safety, security, reliability, and performance of the systems designed by their institutions. These letters are submitted to the Department of Energy, where a cover letter from the secretary is attached for transmission to the president. The secretary of energy can say what he likes in his own letter, but he is not permitted to change any of the input from the laboratory directors, thus ensuring that an accurate technical assessment of the nuclear stockpile is provided directly to the only person who can authorize its use.

As a check on the technical judgment and integrity of the laboratories, U.S. Strategic Command performs its own review of each weapons type. Its Stockpile Assessment Team (SAT), composed of retired weapons experts, some of whom designed and tested weapons themselves, conducts one of the most rigorous scientific examinations that I have ever encountered. Many "reviews" actually consist of prepared presentations intended to convince listeners of the speaker's point, but the SAT digs into details, asks demanding questions, and sometimes assigns homework when it is not convinced by what it is shown. It provides its final report on the stockpile to the commander, who can then share it with the president.

Specific technical topics, such as how long a plutonium "pit" might last in a stockpiled weapon, are analyzed by other organizations, including the Defense Science Board and the National Academy of Science, and ad hoc groups formed as appropriate. All these studies are intended to bring fresh eyes to the assertions of weapons experts, but all have the shortfall of only reviewing what is presented to them. This differs from normal scientific peer review, where technical papers are sent to other scientists who have personal experience doing the type of work reported in the paper and who, at least in principle, could independently reproduce the results. New or provocative results are not generally accepted by the scientific community until someone else has independently verified them. One of the limiting aspects of tight security in the nuclear weapons program is that only a few dozen people are expert on nuclear weapons safety and performance, so it is difficult to arrange for a truly independent assessment.

ALL THE NUCLEAR weapons in America's arsenal were developed through a process of design and testing. Today we face the prospect of maintaining them well beyond their design lifetimes without the opportunity to test them, a major scientific and engineering challenge. Consider the following analogy: Suppose that you lived on an isolated farm and that a family member suffered from a potentially fatal medical condition. The hospital is many miles away and you have several cars in your garage, but under an agreement with your neighbors you have promised never to drive them or even to start their engines except in dire emergency. How can you be certain that these vehicles will work when you need them, when the life of a loved one hangs in the balance? You could check that they are full of gasoline, that their oil is at the proper level, and that their tires are inflated. But suppose that during one of your routine inspections you find that the spark plugs are starting to corrode, raising doubts about the performance of the engine. Worse, the auto supply store tells you that they don't make that type of spark plug anymore. Could you modify one of the newer types to fit? But how would you know that the new spark plugs really work without actually starting up the engine? Would you break your agreement with your neighbors, or would you accept the risk and hope for the best?

This simple analogy gives some insight into the challenge of maintaining our nuclear arsenal in the absence of underground nuclear testing. No one has ever tried to maintain objects as complex as nuclear weapons in perpetuity without testing them in their most vital function, which in the case of a nuclear explosive is its ability to explode when needed and to be safe in all other credible environments. During the Cold War, we created a stockpile of highly sophisticated weapons that were intended to be periodically tested to reveal unforeseen problems that might have arisen from aging, or "birth defects" dating from their original construction. We assumed that no weapon would remain in the stockpile for more than ten or twenty years, so relatively few spare parts were made during the original production run. Less consideration was given to making weapons last many decades than to making them more capable in their assigned missions.

The decision to stop testing was a combination of politics—the hope of impeding proliferation by preventing nuclear aspirants from testing *their* weapons—and an assessment of the technical risk of not testing our own weapons. President George H. W. Bush concluded that the potential benefits of not testing outweighed the risks and ordered, on October 1, 1992, a temporary moratorium on nuclear testing by the United States. President Clinton continued the moratorium and signed the Comprehensive Test Ban Treaty, which would have made the testing prohibition permanent, but the Senate refused to provide its consent; as a result, the United States was not obligated by the president's signature. Congress did, however, endorse the de facto test moratorium by withholding funding for the nuclear test program, essentially making it impossible for the administration to conduct a test even if it wanted to, but it did not go so far as to make it against the law to test.

All the nuclear weapons in our arsenal are either beyond their intended design lifetime or rapidly approaching that point. Disassembly of weapons for diagnostic checks has revealed a number of problems that were never anticipated, and some of these findings have required considerable effort to understand and fix. Dealing with them has been complicated by the fact that the nuclear weapons production complex in the United States has all but collapsed, making some parts impossible to replace and others replaceable only by similar, but not quite identical, new parts. How did this happen, and what is the prospect for maintaining our nuclear arsenal without the ability to manufacture and test replacement parts?

THE EUPHORIA THAT accompanied the end of the Cold War was in full swing when nuclear testing ended in 1992. There was an almost audible sigh of relief that the threat of nuclear Armageddon, one that had hung over the world since the dawn of the Cold War, was receding. There was a hope that much of the money formerly allocated to nuclear weapons would be shifted to solve urgent social problems such as health care, education, and economic development. Since there were no requirements for new weapons and since testing had been halted, Congress began a series of budget reductions in the nuclear weapons program that, if carried on at a constant rate, would have resulted in zero funding in 1999. Each year the laboratories and plants jettisoned capability to balance the books, capability that they might never be able to replace given changing laws that protected the environment and worker safety. Some people applauded the reductions, believing them a route to nuclear disarmament. Others feared that the world was changing too fast and too unpredictably to risk giving up our deterrent; they fought for maintaining at least some capability in the nuclear weapons laboratories and production plants.

The Department of Energy tried to plot a middle course between the two extremes. Recognizing that there was simply not enough money to go around, DOE gave priority to preserving the intellectual capital in the nuclear weapons design laboratories, the people who understood why weapons were the way they were and how to design new ones if the need arose.

What suffered in DOE's strategy were the production plants that, it reasoned, were not producing any new weapons at the time and that could be fixed later if required. Plutonium production facilities at Richland, Washington, and Savannah River, South Carolina, were put into standby status. The nation's only plutonium pit fabrication facility at Rocky Flats, Colorado, was already closed (it had been raided by the FBI in 1989 for violations of environmental laws), and rather than refurbishing this aged plant, it was converted to an environmental research park with the eventual goal of cleaning it up and closing it for good.

This was accomplished in 2005, ahead of schedule and well under budget. The Y12 uranium fabrication plant at Oak Ridge, Tennessee, lost thousands of workers as its focus turned from weapons production to environmental cleanup and the conversion of its once super-secret production lines to other, nonnuclear, industrial tasks. Collapsing budgets forced these decisions to be made so rapidly that there was not even time to mothball equipment—it was salvaged, destroyed, or simply left to rust for lack of funds to maintain it. Only the Pantex Plant near Amarillo, Texas, seemed to have a mission, that being the dismantling of thousands of unneeded weapons.

Deep budget cuts at the nuclear weapons production plants were designed to protect the three nuclear weapons laboratories, two focusing on the physics design of bombs and warheads and one focusing on engineering. Los Alamos National Laboratory, in New Mexico, was the original design laboratory during the Manhattan Project and grew into a multifunction scientific institution employing more than nine thousand people. Spread

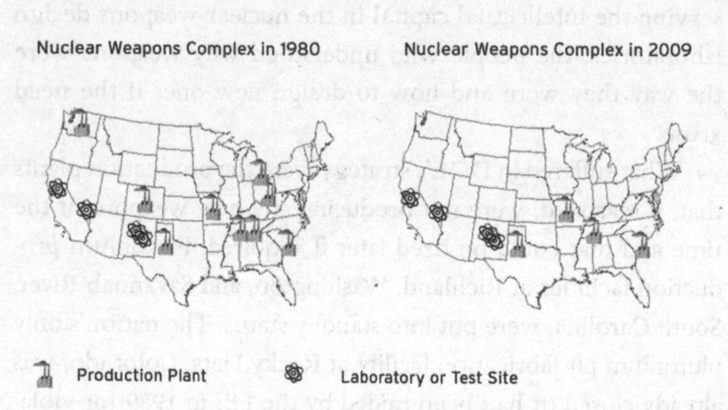

Comparison of the nuclear weapons production complex in 1980 and 2009. Several sites have already been closed and further consolidation is planned.

over forty-three square miles, Los Alamos is known for outstanding science and has extensive capabilities for experiments involving nuclear materials and high explosives. Los Alamos scientists and engineers designed most of the nuclear weapons in the U.S. stockpile.

The second physics design laboratory is Lawrence Livermore National Laboratory, located east of San Francisco. Situated in a compact one-square-mile area, Livermore boasts the fastest computers in the world and is finishing construction of the National Ignition Facility, a mammoth laser designed to produce thermonuclear fusion on a small scale.

Sandia National Laboratories has its main operations in Albuquerque, New Mexico, and a second facility in Livermore. Sandia does all the engineering work associated with the nonnuclear components of weapons—electronics, structural and mechanical parts, and security systems. It has a wide array of capabilities, including machines that can generate intense gamma ray and neutron pulses to verify that electronic and other components will continue to work in the radiation cloud near a nuclear blast.

Even though the United States was not designing any new nuclear weapons, DOE still thought it essential to maintain two physics design laboratories. Nuclear weapons were too complex, and the danger of overconfidence was too great, to place total reliance on one design group. Los Alamos and Livermore were well known for their intense competition—it was sometimes said that they designed their weapons with each other in mind. Competing designs were held in the strictest secrecy—not just from the Soviets, but from the other laboratory. A "separate but equal" attitude existed and occasionally was carried to an absurd degree. For example, at the Nevada Test Site where both labs performed nuclear tests, each labora-

The National Ignition Facility (NIF), the world's largest laser, at the Lawrence Livermore National Laboratory. This stadium-sized laser can simulate some conditions at the lower end of nuclear weapons performance.

tory had its own dormitories, automobiles, test areas, tunnels, and even radio frequencies. Given the intensity of their rivalry, DOE was sure that they would be critical of each other's assessments of weapons status. Sandia could continue as the sole engineering laboratory since most of the components that it produced—those outside the nuclear explosive package itself—could be tested in nonnuclear experiments.

A key problem in maintaining capability at the laboratories was giving the personnel, many of whom had unique skills developed during decades of weapons development and testing, something useful and interesting to do. There was a very real concern that universities and industry would see the well-trained staff at the labs as easy pickings and that a "brain drain" would erode the nation's nuclear weapons expertise. This situation was not unlike similar challenges faced by the navy or air force when they had to maintain unique submarine yards or aircraft design capabilities between major weapons systems acquisitions. DOE adopted what it called an "anchor store" approach ("Every shopping mall needs an anchor store.") by funding one major facility at each laboratory—a large laser at Livermore, a neutron-scattering facility (for materials science) at Los Alamos, and a microelectronics research facility at Sandia—that were intended to retain the best people and recruit the next generation of weapons scientists and engineers. Each laboratory was provided with the most powerful computers in the world, machines capable of performing more than one hundred trillion operations per second, about the equivalent of one hundred thousand personal computers working in unison.

All these facilities are part of the Stockpile Stewardship Program that is designed to maintain the nuclear stockpile into the indefinite future without nuclear testing. By careful inspection of disassembled weapons and the analysis of old nuclear test data, scientists hope that they can identify all the essential elements of weapons safety and performance and address each one of them by means of computer calculations and nonnuclear experiments. For example, if you are concerned about tiny cracks in high explosive affecting the quality of a weapon's implosion, you could disassemble the weapon, replace the plutonium with a nonnuclear material such as lead, and explode the modified unit. Scientists believe that if a computer code accurately predicts the implosion of the lead substitute, it should do as well for the real weapon. DOE approved construction of a giant x-ray machine at Los Alamos, the Dual Axis Radiographic Hydrotest Facility, to perform these tests. It uses a series of intense x-ray pulses—many millions of times more powerful than those used in medical x-rays—to peer inside a weapon while the implosion is in progress.

The advances made by the weapons laboratories in computational physics and weapons-related experiments are impressive, but several concerns remain about the viability of the stewardship program to maintain weapons forever without testing them. Most important, some key weapons processes simply can't be duplicated in the laboratory since the energy densities achieved in a functioning nuclear weapon are higher than anything in nature short of a supernova or a black hole. Not even the stadium-sized laser at the National Ignition Facility can reproduce the extreme temperatures and pressures in a nuclear explosion. Scientists must extrapolate from what they can measure in the laboratory to what actually happens during a nuclear explosion.

Another assumption in stockpile stewardship is that one can study each piece of a nuclear explosion in isolation and then use computers to glue them back together into a prediction for a real weapon. Science has long used the principle of reductionism, or the taking apart of complex things to understand their basic elements, but it is less certain that the process can be reversed and that the pieces can be put back together to yield the original object. In some cases, scientists have found that it is *impossible* to model things as well as we would like, including the weather. We are getting quite good at modeling weather over periods of hours or days, but we now know that the chaotic nature of the atmosphere makes it *theoretically impossible* to make accurate predictions of the weather many months in advance, no matter how powerful our computers or how clever our programmers. Nuclear weapons do not appear to be as difficult to model as the weather, but they are sufficiently complex to raise concerns that, with no ability to test the results of our computer predictions, we can be sure of our answers. It was just this concern that led the U.S. Senate to demur from ratification of the Comprehensive Test Ban Treaty.

Perhaps the greatest potential danger of the Stockpile Stewardship Program is overconfidence based on lack of hands-on experience. It is not uncommon to hear nuclear weapons designers assert that "we can do that without testing" for concepts that formerly would have undergone an extensive *series* of underground nuclear tests. Are we so much smarter today than we were at the end of testing, or are we a victim of what Kenneth Johnston, former chief scientist of the Atomic Weapons Establishment in England, said happened during the UK's moratorium on nuclear testing: "Our confidence went up as our competence went down."

I no longer believe that we can maintain our existing arsenal of nuclear weapons indefinitely without nuclear testing. That does not mean that I advocate returning to testing, but only that I don't think that we can continue on our *present course* and still have high confidence in the safety, reliability, and performance of the nuclear arsenal. We have several options: First, we could accept lower confidence in individual weapons and compensate for it by assigning more than one weapon, and perhaps more than one type of weapon, to each target, akin to keeping more than one car in the driveway. This works so long as we are confident that at least *some* weapons will work, but it necessitates keeping a much bigger stockpile than we need. Second, we could introduce a new class of nuclear weapon that is simpler and more rugged than the ones that we have now. This would require confidence that these untested weapons were more reliable than our older but tested designs. Third, we could simply allow our weapons to decay and our confidence to erode, our only comfort being that all the other nuclear states would be in the same predicament. But the United States already has the world's oldest stockpile and no manufacturing capability for some key components, putting us at a potential disadvantage compared to other nations. Fourth, we could accept the political fallout resulting from an occasional nuclear test. There is no perfect answer to this question since each of the options has pros and cons.

My skepticism regarding the ability of the Stockpile Stewardship Program to maintain the arsenal indefinitely without nuclear testing is not shared by all my colleagues. Some believe that a method referred to as "Quantification of Margins and Uncertainties" can help us estimate the accuracy of our predictions and enable us to steer clear of performance problems. The idea underlying this method seems obvious: Find out what parameters in a weapon introduce the greatest risk to its safety and performance, and back off on those parameters so that the risk goes down. However, the notion that we can back away from a performance cliff assumes that it is indeed a *cliff* and not a *table*, where backing away from one edge only leads you closer to another edge. Not knowing exactly where the edges are makes it difficult to put yourself in the center of the table—the place with the least risk. Computer simulations can only describe those phenomena that are coded into them, and on more than one occasion we have been surprised when confirmatory experiments gave results different from our calculations.

A couple of examples might help to illustrate the danger of missing physics in a computer code. High explosive, vital to the performance of a nuclear weapon, is actually a mixture of explosive granules and a plasticizer that gives it mechanical strength and enables it to be machined to precise shapes. Over time this plasticizer evaporates, making the high explosive brittle and prone to cracking. The same thing happens to the plastic dashboard of an automobile when, after many years, the plasticizer evaporates and cracks appear. But while cracks in a dashboard are only cosmetic, cracks in high explosive can affect safety, raising serious questions about the handling and storage of weapons. Most computer codes model high explosive as a smooth and continuous solid and have no ability to model microscopic cracks. If you just looked at the results of computer calculations and did not look at the real explosive, you would never know there was a problem.

In 1994, several scientists proposed to use a large laser at Livermore to look at a simple process that had important implications for weapons performance. Some of their colleagues criticized them for wasting money. "If we know *anything* about nuclear weapons, we know *that*," was one comment. When the experiment was performed it gave exactly the *opposite* result from the computer calculations—a puzzle to the scientists until they realized that a different process than the one that they anticipated—a process not included in the computer code—was responsible for the observed results. Subsequent calculations using the correct physics were able to model the experimental results, but this episode made some people nervous about using the phrase, "We can calculate that."

Surprises happened in nuclear tests. One of my designs performed so unexpectedly that the recording equipment had trouble capturing the data. In designing it I ran every computer code available. Scientists at the other weapons laboratories ran their own codes and got the same results as I did. Extraordinary care was taken in the construction of the device, and numerous technical reviews were held to critique its performance. After all this diligence, nature had other ideas. I still remember with crystal clarity my division leader saying, at the end of a particularly detailed review, "There's nothing wrong with this bomb." There wasn't, in fact—it just worked differently than we predicted. We eventually figured out what happened—it was an error in a data table—but no one found the problem before the device was tested.

Science has always proceeded through a comparison of theory and experiment. Lacking the ability to conduct nuclear tests, we will never be completely sure that we have included everything in our calculations. Many Nobel Prizes have been won by scientists doing experiments that "don't need to be done." Computer simulations are just that—imitations of the real thing—and I know of no way to be *sure* that a calculation produces an exact rendition of what happens in nature.

THE DECISION TO halt nuclear weapons testing had a strong political component. Some claimed that by not testing, the United States would discourage other countries from testing. The risk of our own weapons developing a problem was seen as lower than the risk associated with other countries modernizing their own stockpiles. Also, by not testing, we would pressure potential proliferants to refrain from testing their first designs, raising doubts as to whether they would work.

Russia, China, Great Britain, and France have also forsworn nuclear testing; each has its own approach to maintaining its nuclear forces. Russia has a program of laboratory experiments and computer simulations similar to that of the United States. It lacks some of the supercomputer technology employed by the U.S. program, but it compensates for this by the high quality of its scientists ("You compute, we think," a Russian scientist once said to me) and by conducting an extensive series of experiments at its former nuclear testing range at Novaya Zemlya, north of the Arctic Circle. In the early years of the twenty-first century the Russians dramatically increased the pace of activity at Novaya Zemlya, even operating during the winter months when complete darkness prevails and attack by polar bears is an ever-present danger. Satellite photographs on GoogleEarth reveal new buildings, a school, and miles of cables leading into experimental areas located deep under mountains.

The United Kingdom, France, and China have smaller stewardship programs than the United States and Russia, but sufficient ones to maintain their smaller stockpiles. All rely on laboratory experiments plus computer simulations to assess the status of their weapons, and all have built large lasers and other facilities to compensate for the end of nuclear testing.

The United States differs from all other nuclear powers in that we are the only one that does not regularly remanufacture and replace our weapons. All other countries believe that the maximum shelf life of a nuclear explosive is in the range of ten to fifteen years. Beyond that they worry that corrosion, cracks, and other age-related problems mandate replacement of the entire weapon. The United States believes that quality control in the initial production of our weapons, along with improved understanding of fundamental phenomena made possible by stockpile stewardship, will enable our weapons to last many decades.

The United States could not emulate the remanufacturing strategies of the other nuclear nations even if we wanted to, since we lack key capabilities to make essential parts of weapons. DOE knew that it was dealing the nuclear plants a near-fatal blow in the mid-1990s by diverting resources to the laboratories, but it did so with the idea that at some point we could resurrect those capabilities, some of which resided in the nonnuclear industrial complex. "We'll fix it later," was the phrase used in many discussions. As the years have passed and more unforeseen problems have occurred in our weapons, we can say that "later" has arrived.

Of the three nuclear materials found in a modern weapon—uranium, plutonium, and tritium—we currently have a limited capability to manufacture two: uranium and tritium. There is a small plutonium capability at Los Alamos, but it is inadequate to deal with a situation that would necessitate replacement of hundreds of plutonium pits in a short period of time. Several starts have been made on plans for a new plutonium facility, each of them sunk by Congress, which refuses to commit substantial funds without a better estimate of just how many weapons we need, of what types, and how big the proposed plant would need to be to meet those needs. Since a plutonium facility is likely to cost more than $10 billion and take more than a decade to complete (three times longer than it took to make the first atomic bomb, including the construction of a national complex of laboratories and production plants), Congress has good reason to ask for a detailed justification of the expenditure. The blank checks that the country wrote to support the nuclear stockpile ended with the Cold War.

A further complication affecting plans for plutonium fabrication is that weapons scientists are uncertain about how long plutonium components will last and hence when we will need to replace them. Do we need to remanufacture all our weapons within a few years, or can we afford to spread out production over a decade or more, permitting us to construct a much smaller and less expensive plant? The natural radioactivity of plutonium causes damage to the metal at the microscopic level, damage that may—or may not—affect the performance of the weapon. Plutonium was first made in 1943, so our experience with its aging is poor compared to how long we anticipate

The Pantex facility, which assembles and disassembles nuclear weapons, near Amarillo, Texas.

keeping weapons in the stockpile. The weapons laboratories have done a few experiments to simulate the aging of plutonium, but the key word is "simulate." Did the method used to accelerate aging give an accurate prediction of what happens in the real material? Knowing how long a plutonium pit will last is an essential data point in designing a new production plant. I do not believe that we have enough data to answer this question.

Uranium components for nuclear weapons are manufactured in the Y12 plant in Oak Ridge, Tennessee. DOE has invested hundreds of millions of dollars to bring this Second World War–era facility back into operation. Tritium, an isotope of hydrogen and an essential ingredient in modern weapons, is stored at the Savannah River Site near Aiken, South Carolina. While essential to weapons, both uranium and tritium present far fewer challenges to scientists and engineers than plutonium.

Sometimes the problems in weapons remanufacture are associated with much more mundane materials—plastics, rubber, and metals such as beryllium. Changes in environmental law and the discovery that some of these materials might cause cancer or other diseases have led to a prohibition against their future use, even in nuclear weapons that contain far more dangerous substances. These laws mean that we are not permitted to "make it just like we made it before." But even if we *could* get the same materials, it is hopeless to try duplicating what was done decades ago, something illustrated in the automobile analogy presented at the beginning of this chapter. A few moments of thought will reveal that we really couldn't remanufacture a 1968 Ford *exactly* the way it was done back then without a massive investment in antiquated equipment that no modern machinist would think safe to operate. We would substitute one plastic for another, use different alloys of steel, and have to accept slightly different formulations of rubber. The same thing is true of a nuclear weapon manufactured in 1968. The cumulative effect of many small changes is one of designers' greatest worries when considering the refurbishment of old nuclear weapons.

THE UNITED STATES has not ratified the Comprehensive Test Ban Treaty, so there is nothing preventing us from resuming underground nuclear testing should the president decide that it is in our national interest. Such an experiment would be done at the Nevada Test Site, located about sixty miles northwest of Las Vegas. Surrounded by Nellis Air Force Base and other federal land, the Nevada Test Site is arguably the best location in the world for conducting nuclear tests owing to its very deep water table, its geology of dry rock and sand, and its isolation from populated areas. Underground experiments would be conducted six hundred to two thousand feet below the desert floor so no radioactivity would be released into the environment.

Although nuclear tests stopped in 1992, the Nevada Test Site is still very busy supporting the weapons program. The isolation and high security of the site make it the preferred test bed for experiments involving plutonium, some of which are conducted in a laboratory complex nearly one thousand feet

A subcritical experiment being prepared to be conducted one thousand feet underground at the Nevada Test Site. The purpose of this experiment was to measure the basic properties of plutonium.

underground. These studies, called subcritical tests because they do not produce any nuclear yield, involve state-of-the-art diagnostics to gather data to test computer codes. They reduce the uncertainty associated with our assessments of weapons and occasionally reveal things that we didn't know about the unique materials in nuclear explosives.

Estimates vary as to how long it would take the United States to resume nuclear testing. If the goal were simply to demonstrate that a weapon will explode, perhaps to show political resolve after other countries resumed testing, then an experiment could be conducted in as little as six to nine months. A more sophisticated experiment involving new measurement apparatus might take two years or more to develop and field. The United States has no plans to conduct a nuclear test, and funding for our readiness program is inadequate to maintain our capability. This is in stark contrast to the Russian test readiness program, which, I am told by Russian colleagues, is very well funded with the construction of new facilities and the development of new capabilities.

THE NATIONAL IGNITION FACILITY, the x-ray machine at Los Alamos, and refurbishments at the nuclear weapons plants are multibillion-dollar investments in America's nuclear arsenal. But hardware alone will not provide the required confidence in our future stockpile. *People* design, maintain, and manufacture weapons. A major worry in the nuclear weapons community is how we recruit and train the next generation of weapons scientists and engineers. Fortunately, the technical qualifications of people who are willing to undergo the rigorous security investigations associated with a "Q," or nuclear weapons clearance, have remained high during the course of the stewardship program. The scientific challenges, the opportunity to work with some of the most advanced equipment in the world, and the patriotism of young Americans have attracted a number of young people into a nuclear weapons career path. These new entrants will never be able to get the nuclear test experience of their older colleagues, so there is an urgent need to pair them with mentors who can guide and temper their enthusiasm for undertaking risks.

SO FAR WE have focused on the problem of maintaining our nuclear explosives. A similar challenge applies to the delivery vehicles that carry them. Missiles, bombers, and submarines are complex machines with finite lives, often designed and constructed by special teams not unlike those who worked on nuclear explosives. Unlike nuclear explosives, however, all our delivery vehicles can be tested, often under highly realistic conditions. Bombers are regularly exercised by loading and launching simulated nuclear bombs and cruise missiles. Each year the United States, Russia, and some other nuclear nations conduct ballistic missile tests in which an ICBM (land-based missile) or SLBM (submarine-based missile) is launched with dummy warheads replacing its nuclear payload. Extensive telemetry and radar coverage reports the detailed flight characteristics of the missile to assess its performance.

Our nuclear weapons delivery systems are aging—the B52 bomber first flew in 1954, and, while it has been heavily modified to keep up with the times, it is an old airplane. More serious is the fact that missiles can be flown only once—they cannot be recovered, refueled, and placed back in their silos for future use. Since only a finite number of Minuteman III missiles were produced, each one tested means one fewer in the

Test launch of a Minuteman III ICBM from Vandenberg Air Force Base in California.

operational force. The air force is actively debating the tradeoff between more tests and more missiles. With the agreement to reduce the number of strategic nuclear weapons from many thousands to between seventeen hundred and twenty-two hundred, we have some missiles to spare, but over time these too will be used up in tests. Arms reduction agreements have delayed but not eliminated the eventual need to build new missiles and bombers.

NUCLEAR WEAPONS PRODUCE unique effects such as electromagnetic pulses, intense radiation, and super-strong shock waves. We need to maintain our understanding of these effects so that we don't suffer catastrophic failure of critical systems in the event of even a limited nuclear attack. The sensitivity of microelectronics to electrical interference, combined with our nearly complete reliance on computers to manage our defense, demands that we have a *better* understanding of nuclear weapons effects than ever before. Yet these capabilities have all but disappeared. There used to be several experimental facilities to simulate the effect of a nuclear electromagnetic pulse on major systems such as aircraft, tanks, and commercial power networks, but today there are few to none, depending on funding in an individual year. Most of the expertise in nuclear weapons effects resides in the for-profit defense sector, and companies are reluctant to maintain expensive facilities in the absence of paying customers. Fixing this problem is not expensive on the scale of major defense programs—an increase of $40 million per year would be sufficient—but so far every attempt to secure even this modest level of funding has failed. Soon we will no longer be able to offer a credible answer to the question, "What would happen if a nuclear weapon exploded nearby?"

9

The Role of Nuclear Weapons in the Twenty-first Century

Most discussions of nuclear weapons start from one of two polar opposite assumptions—either they are a liability and should be eliminated, or they are essential and must be maintained. Some view nuclear weapons as inherently immoral, unnecessarily destructive of civilian populations, and unusable in any "just war." Others see them as bulwarks against evil, an assurance that the world wars that decimated the twentieth century will not blight the twenty-first. Coming from opposite directions, and lacking much in the way of factual information with which to frame a debate, it is little wonder that American nuclear policy has seen scant change since the end of the Cold War. But nuclear weapons are too important to let our policy drift, carried by the momentum of a different era. If we elect to continue to have them, we must understand why, how many we need, and the purpose we intend for them. If we elect to eliminate them, we should understand the challenges and the risks that will follow.

I strongly believe that a meaningful discussion of the future of nuclear weapons must start by considering the geopolitical context in which they exist and the military requirements they are intended to meet. Is deterrence still a valid concept since the end of the superpower standoff, or should we create a new doctrine, perhaps one based on the precision use of low-yield nuclear weapons for tactical applications? Is there any place at all for countervalue strategies that target cities as a means of striking fear in the hearts of potential aggressors, or should we continue a counterforce strategy that targets only military capabilities posing an imminent threat to the United States? On a more fundamental level, do we need nuclear weapons at all at a time when our country has a substantial conventional military advantage?

NUCLEAR WEAPONS PLAYED a key role in the Cold War standoff between the capitalist West and the communist East. Each side saw them as the ultimate insurance that it could not be defeated, at least militarily, and each saw them as cost-effective force multipliers on the battlefield—one nuclear weapon could substitute for thousands of troops. Nuclear weapons served both strategic and tactical roles, the former played out upon the broad stage of international relations and the latter working at the level of the tank, ship, or bomber. Today we face a fundamentally different set of world conditions that play no less significant a role in shaping our thinking about nuclear weapons than did the superpower standoff.

The Soviet Union, the principal driver for the development of

American strategic weapons, is gone, but the Russian Federation is modernizing its massive and capable nuclear arsenal. Russia is the only country in the world with a sufficient nuclear force to utterly destroy America—the mere existence of these weapons demands that we factor Russia into our nuclear strategy.

Some commentators discount the future of Russia, citing the huge problems the country faces in reforming a dilapidated industrial plant and in dealing with a shrinking population beset by poor health, emigration of its best and brightest to other countries, and government corruption. While their reasoning is compelling, it is also true that this unique country has been in worse shape and recovered rapidly. Russia was in ruins at the end of the First World War and the Revolution. Millions had died and the national infrastructure had nearly collapsed. However, within a generation Russia was back on its feet and boasted one of the world's largest air fleets, a massive army, and an impressive university system. More than twenty million Soviets perished in the Second World War, many at the hands of the Stalin regime. European Russia, home of much of the country's industry, was again in ruins. But the Soviet Union detonated an atomic bomb just four years after the end of the war, and it launched the world's first artificial satellite in 1957—about a half generation after the war ended. Russia suffered a series of catastrophes in the twentieth century, but it recovered from each of them in less than a generation. We are now nearly a full generation past the breakup of the Soviet Union.

Current trends in Russia, which are remarkably predictable, suggest that the country will reemerge as a significant military power. Russia has periodically flirted with Western-style reform, only to experience problems that drove it back to a strong central government. President Vladimir Putin, the most capable and energetic Russian leader in many decades, reversed a liberal trend toward decentralization and consolidated power in the Kremlin. No longer does one hear talk of the oblasts (local states) becoming "semi-autonomous regions." Governors are appointed by, and directed from, Moscow. The government has taken control of most of the press, has placed stringent new laws on the operation of foreign organizations on Russian soil, and has reasserted its historical ascendancy over the armed forces.

Russia recognizes the inferiority of its conventional military forces and, just as NATO did during the Cold War, is using its nuclear weapons to compensate. It announced a new generation of low-yield weapons for use on accurate short-range ballistic missiles placed around the country's borders, a clear signal to anyone who would covet resource-rich areas in the South and East. On the strategic front, Russia is modernizing its ballistic missile fleet with the SS–27 ICBM. There are hints that the single warhead on the SS–27 is capable of maneuvering to avoid ballistic missile defenses, another signal that Moscow is not to be trifled with in the nuclear arena.

Russia is no longer considered a strategic adversary of the United States, but it must still figure prominently in any discussion of our nuclear arsenal. One of the chief functions of nuclear weapons is to counter any potential threat on the strategic scale, no matter how unlikely it may now appear. Another is to *prevent* threats from arising by making the risk too great for the potential aggressor. Factoring Russia into our future nuclear requirements is not an argument for restarting the arms race, a wasteful and unnecessary act on either side's part. It simply recognizes that Russia continues to maintain and modernize one of the largest strategic nuclear stockpiles in the world. Until and unless Russia reverses the expansion of its nuclear arsenal, it would be foolish to ignore it.

CHINA IS OFTEN mentioned as a future peer competitor of the United States. Some generals are convinced that war between America and China is inevitable—a chilling echo of General Curtis LeMay's recommendation for a massive preemptive strike on the Soviet Union. China's astonishing economic growth and its continuing threat to take Taiwan by force are two principal concerns. Construction in major Chinese cities is so rapid that it is placing strains on the world's ability to supply construction materials. China consumes 40 percent of the world's concrete, and its appetite for dwindling reserves of oil is growing faster than that of any other country. The list of Chinese millionaires seems to grow daily. Some experts predict that China's economy could outstrip that of the United States within twenty years.

China, with its population of more than one billion people, can exert influence not only as a provider and consumer of goods and services, but also as a military power. Few other countries could hope to field an army as large as China, and few countries could absorb more punishment in a major war and continue to function.

Linear projections of growth often ignore serious underlying problems, of which China faces many. China is in the midst of a demographic crisis, ironically one arising from the success of its birth control programs. Concerned about its ability to feed a massive and growing population, China limited households to one child, with sanctions against couples who had more. This policy, combined with the traditional Chinese preference for male offspring, resulted in a fundamental imbalance in age and sex distribution, with prospects for a large class of older people reliant on a small number of young men for support. The country lacks an effective social security system, so a young Chinese man may find himself responsible for supporting six other people, including his wife, child, parents, and in-laws—not to mention any other relations who do not have male children to help them. Such a burden restricts the ability of a young family to save and to buy durable goods supplied by domestic industry.

Perhaps even more challenging is that the majority of Chinese live in the countryside, with poor medical care and education. These rural peasants are demanding services comparable to what they see in cities such as Beijing and Shanghai, and the central government, still at least nominally communist, clearly understands that these demands cannot continue to go unmet without serious trouble.

China boasts massive armed forces, but closer inspection reveals that the quality of its equipment is well behind that of the United States, Russia, and other developed countries. The Gulf wars demonstrated that numbers alone are no match for technologically superior forces. The People's Liberation Army and Navy learned this lesson and embarked on a program of modernizing everything from soldiers' rifles to submarines.

With only a few dozen missiles capable of reaching the United States, each of which is said to be equipped with a multimegaton warhead, China appears to be squarely in the countervalue targeting camp; that is, its weapons are pointed at cities rather than military targets. One might assume that China considers its nuclear arsenal an insurance policy against major strategic aggression, not an instrument of preemptive war. Objections to America's deployment of national missile defense led China to threaten to increase its number of deployed missiles and to equip them with a more accurate and improved warhead, but the speed with which they will pursue this modernization remains to be seen.

I think that China will remain focused on internal and economic issues for some time to come, perhaps for a generation or more. Beijing is developing diplomatic and trade relationships around the world, including in developing Africa and other resource-rich areas. It appears to have little appetite for aggression that could put it at odds with America, a huge consumer of Chinese manufactured goods. Periodic bursts of rhetoric over Taiwan are to be expected, given strong historical feelings about unification, but one must question whether Beijing wants the island enough to risk a major war that would return it to a position of isolation after years of trying to integrate itself into the global community.

THE PROLIFERATION OF nuclear, chemical, and biological weapons has increased over the past two decades, with India, Pakistan, and North Korea testing nuclear weapons and Iran apparently pursuing them. In each of these cases, one could argue that the country involved saw nuclear development more as a means of assuring national sovereignty than as a threat to the United States or other Western powers. Their arsenals are small and their means for delivering weapons are limited, although the rapid proliferation of ballistic missile technology may soon give even small countries the ability to deliver nuclear weapons over intercontinental distances. No developing country is likely to represent a strategic challenge to the United States anytime soon, and all of them recognize that the conventional military forces of the United States are capable of inflicting major damage even without resort to nuclear weapons.

Nuclear weapons are no longer needed to deter a massive Soviet attack on the United States or an invasion of Western Europe. However, just as they did during the Cold War, they maintain a strategic balance, assuring each side that it cannot hope to prevail over the other. This balance has spread beyond the dyad of the United States and Russia—other nations recognize America's conventional military supremacy and have decided to weather international condemnation to build the only type of weapon that could defeat us on the battlefield.

IN CHAPTER 5, we discussed new technologies that can perform missions previously assigned to nuclear weapons—without incurring the huge political cost associated with crossing the nuclear threshold. The United States Strategic Command has been integrating some of these ideas into its strategic war plans, reducing the need for nuclear weapons while achieving comparable levels of confidence in the outcome of any engagement.

However, there are still some cases where nothing short of a nuclear weapon will do the job. WMD facilities located deep underground, or in tunnels bored into mountains, cannot be defeated by conventional weapons no matter how precisely they are guided to the target. Only the explosive force of a nuclear weapon will destroy them, and only by keeping a sufficiently capable nuclear arsenal can we warn potential adversaries that even their best efforts at hardening will ultimately fail.

There is another aspect of nuclear weapons policy that gets much less attention today: the threat of attacking cities, industrial centers, and cultural assets, the countervalue strategy that we have repeatedly tried to distance ourselves from due to its enormous cost in human life. I hope that the United States will continue to avoid purposeful targeting of civilian populations, but it is naïve to ignore countervalue targeting as a component of deterrence. Enemies will always know that the United States has the capability to destroy their cities, and they will factor this into their decision calculus. Other countries, particularly those with only a few nuclear weapons, are likely to continue to employ countervalue strategies as the core of their nuclear deterrent posture for the simple reason that they lack the numbers and types of weapons to implement a counterforce approach. The subtleties of policy aside, nuclear weapons are blunt instruments of warfare, icons of mass destruction.

NUCLEAR STRATEGY IS complex because it must deal with uncertain futures—both the geopolitical context in which such weapons will exist and a constantly evolving set of military technologies both here and abroad. Before beginning a discussion of alternate nuclear futures, I would like to propose six basic considerations that could frame a debate on the role of nuclear weapons in the twenty-first century.

- ***The fundamental role of nuclear weapons is to maintain strategic peace.*** I know of no military commander who anticipates the need to use nuclear weapons in any case short of the survival of the nation. We no longer need nuclear weapons to compensate for quantitative or qualitative weaknesses in our conventional forces. We should emphasize that these are *strategic* weapons, the ultimate guarantors of the survival of our country against state-level aggression. We should work toward the elimination of tactical nuclear weapons and insist that other countries do likewise, raising the threshold for any possible use of these uniquely destructive weapons.

• ***The United States should be among the last nations to use nuclear weapons.*** Our use of a nuclear weapon could justify another country in using one against us. Only when we, or an ally, have already been struck by a weapon of mass destruction, or when we are absolutely convinced that such a strike is imminent, should we consider the use of nuclear weapons. The United States should constantly emphasize that nuclear weapons are qualitatively different from other types of weapons and that *any* nuclear use crosses a fundamental threshold in international relations. Our conventional military superiority means that we can win any battle without resorting to nuclear weapons.

• ***The United States should adopt a policy of purposeful ambiguity regarding the first use of nuclear weapons.*** To state outright that we will not be the first to use nuclear weapons sends a clear signal to a potential adversary that it can be the first to throw a nuclear punch. There is value in ambiguity in that it complicates the attack plans of a future adversary. We must be careful, however, that we are not *so* ambiguous that we tempt an adversary into launching a first strike against us in fear that we are about to attack.

• ***The assurance of massive destruction is an implicit element of deterrence.*** The United States should continue to employ a counterforce strategy, one that focuses on reducing or eliminating the enemy's ability to inflict damage on our country. But we should remember that other countries will worry about the loss of cities and vital resources. Purposeful ambiguity will raise questions in the minds of potential adversaries, questions that will constrain their actions and, hopefully, prevent them from attacking us. The threat of massive nuclear retaliation—

whether explicitly stated or not—is part of any perception of nuclear weapons.

• ***Nuclear weapons are primarily instruments of state-to-state deterrence.*** Critics of nuclear weapons argue that they have little value in deterring terrorists. I agree with this assessment. Nuclear weapons serve as deterrents to state-level entities that have assets that can be held at risk. They are ineffective when the enemy is not concentrated in a given area or when they lack military assets that, if destroyed, would lessen the threat to the United States.

• ***We should assume that we will never have enough intelligence to confidently assess the capabilities and intents of other nations.*** The record of the intelligence community for predicting or even detecting events is poor. Numerous studies have examined why, but in reality we will never be able to know everything in a world dominated by human behavior. We must learn to live with uncertainty, particularly on the strategic front, and configure our nuclear arsenal so that it can respond to the broadest range of future contingencies.

Are there *any* conditions in which the United States might use nuclear weapons? Are they really no more than an existential deterrent, threatened but never used? There are several situations where the limited use of nuclear weapons could prevent catastrophic damage to the United States. Suppose that North Korea, in an insane demonstration of its military power, were to launch a nuclear weapon against Los Angeles, with bellicose threats that San Francisco is next. Having suffered one attack, with a credible threat of a second, the United States might respond in kind by a limited nuclear attack against North Korean missile locations and military installations.

Suppose that, in a future war, we had unambiguous intelligence that an enemy was about to launch a new type of biological attack against our troops, one for which we had no defense and that could cause hundreds of thousands of casualties. If we knew the location of the enemy missiles, we could incapacitate them via conventional means. But suppose that the enemy's biological weapons storage site was buried so deep that only a nuclear weapon would destroy it?

There are no optimal solutions to either of these problems. Not using nuclear weapons might lead to horrific losses on our side. Using them could lead to more nuclear weapons being launched in the future, perhaps against us. The decision the president makes will depend on the details of the situation. Nuclear weapons provide him or her with an option, an ultimate weapon for use against an ultimate threat.

HAVING BRIEFLY ESTABLISHED the context for nuclear weapons in the coming decades, we can now address the questions posed at the beginning of this chapter, namely how many we need and what types. I divide the discussion into four options: the abolitionist position, the minimalist position, the maximalist position, and the moderate position. Each will be discussed in turn, but special attention will be given the moderate position as I believe that it is the best choice for the United States.

The Abolitionist Position

Every president of the United States since the invention of nuclear weapons has discussed their elimination. The problem is that no proposal has been put forward to enable us to verify that the nuclear forces of every country in the world have been reduced to *zero*. Nuclear explosives are so powerful that only a few could create a decisive strategic advantage, something painfully demonstrated at Hiroshima and Nagasaki. One need only imagine a situation in which America had disarmed, only to be threatened by a country that openly tested a nuclear explosive and claimed that it had more at the ready.

Our inability to be sure that, once we disarm, others have done likewise is the standard argument against the abolition of nuclear weapons. However, advances in nuclear detector technology and the success of international agreements for mutual inspections suggest that we take another look at barriers to the elimination of nuclear forces. Is verification of disarmament really impossible? Is there a technology that, combined with a set of inspections, could provide adequate confidence that no country or group had a nuclear weapon or the means to produce it?

This is not as far-fetched as it might sound. The Open Skies Treaty of 2002 is an agreement among thirty countries, including the United States and Russia, that allows photo-reconnaissance flights across their territory. American crews fly missions along predefined routes in Russia, and Russian crews do the same in the United States. We could extend this treaty to equip low-flying aircraft with radiation detectors that could identify large quantities of nuclear material. A determined cheater could move materials inside a mountain or deep underground, foiling detection, but we might think of other methods to deal with this possibility.

The second argument frequently used against abolition—that an adversary with only a few nuclear weapons could intimidate the United States or other countries—also begs closer scrutiny. A few weapons, even of the multimegaton class, would wreak incredible destruction on a few American cities, but they would not destroy the fabric of our country and they would not eliminate our considerable conventional military capability. The United States, along with other outraged nations of the world, would still be able to wage war and defeat the nuclear aggressor using conventional forces. Nuclear weapons would play a potent role in international blackmail, but they would not carry the day. Moreover, the advent of an effective missile defense system, combined with nuclear detectors at our borders and seaports, would make it more difficult for a potential attacker to deliver one or a few weapons.

An alternative to the total abolition of nuclear weapons is to place a small number of them under international control, something proposed by the Truman administration when America was the sole nuclear power. This approach avoids the dilemma of proving a negative—that no country has or is developing nuclear weapons—but it suffers from the need to empower an international body with using nuclear weapons. What are the criteria for nuclear use? What countries get to decide? Would there be a veto authority? America has traditionally been suspicious of granting power to international bodies, at times fearing entanglement in foreign wars and at other times rejecting the need to submit to any authority beyond its own. And even if America agreed, would North Korea and Iran? Would France? International control of nuclear weapons is not impossible, but it faces huge hurdles that will take some time to resolve.

Nuclear abolition is sometimes promoted based on philosophical or religious grounds that view nuclear weapons as inherently immoral. Such arguments are unconvincing to others who see these weapons as a practical response to a manifestly dangerous world. We must move beyond such unproductive debates. Rather than reject abolition out of hand or naïvely pursue it, we need to conduct a rigorous study of its challenges and potential solutions to those challenges, placing all the facts on the table for a rational discussion.

The Minimalist Position

If it is not realistic to eliminate *all* nuclear weapons, could we at least reduce their numbers so that they no longer represent the threat to human civilization that they did during the Cold War? General Andrew Goodpaster, adviser to President Eisenhower, recommended that we might go as low as one hundred weapons, enough to cause great damage in a counterattack, but not enough to ensure victory in a preemptive strike against another nation. This would put America in a comparable position to Great Britain, France, and China and might encourage Russia to follow a similar path.

Another formulation of the minimalist position suggests separating warheads from missiles to lessen the probability that they would be used in a moment of uncertainty or even anger. We would still retain the weapons, but everyone would be able to see that they were not ready for immediate use, greatly reducing the so-called hair-trigger danger of launch. In this scenario, nuclear-armed submarines would remain in port except in a crisis situation, in which case warheads would be installed and the ships would sail.

The hair-trigger argument for separating warheads and missiles is fundamentally flawed. Nuclear weapons can only be launched and detonated following an order from the president of the United States. Local commanders of missile bases, submarines, or bombers do not have the ability to independently order an attack. Launch authority comes from the president in the form of a multidigit code that is unbreakable by any known or projected means. Codes are contained in sealed containers always at the president's side and are constructed in such a way that no human eyes see them during their production. The launch of a nuclear attack is highly choreographed for the precise purpose of avoiding hasty launch.

The minimalist option ignores several practical considerations. First, there is the problem of how to define "one hundred weapons." Is it the total number of weapons in the stockpile, or only those that are capable of being quickly mated to missiles and aircraft? What about weapons included in the supply chain, those undergoing refurbishment or remanufacture? What would happen if a fatal flaw was discovered in the stockpile, one that rendered it unsafe or ineffective? Should a reserve force of a different design be maintained, one that could be quickly substituted for the ailing weapons?

Second, how should we deploy a small number of weapons? Only on survivable submarine-launched ballistic missiles, as the British do? Should we continue to maintain a triad of land-based ICBMs, submarine-launched SLBMs, and bombers? If more than one delivery option is deemed necessary, how should we best divide the small stockpile among them?

Third, the minimalist position has the same problem of verification faced by the abolitionist position, namely the difficulty of locating and placing under suitable controls *all* nuclear material that a potential adversary could use to construct additional weapons.

Lastly, and perhaps most importantly, a small nuclear stockpile could actually *increase* the probability of nuclear war, tempting an attacker to think that it could destroy all or most of our weapons in a first strike, effectively eliminating our ability to respond. The attacker could then dictate terms while threatening a second strike. Removing warheads from missiles has a similar destabilizing effect, since if, during a crisis, we were to remate them or order our submarines to sail, tensions could escalate to the point of war.

The minimalist position offers the apparent attraction of limiting the (still horrific) destruction that would result from a nuclear war, but it suffers when one looks more closely into the details. It is not clear whether the arbitrary reduction of our weapons to a few hundred would reduce or increase the probability of war.

The Maximalist Position

The maximalist position, still held by some in the nuclear policy community, argues that the United States should maintain a sizable arsenal to respond to any contingency, including the emergence of a nuclear peer competitor. This is especially important, they maintain, during a period when the United States does not have the ability to manufacture new nuclear weapons in any quantity.

Conservative calculations suggest that to maintain an arsenal of two thousand deployed strategic weapons (near the middle of the range of seventeen hundred to twenty-two hundred agreed to by presidents George W. Bush and Vladimir Putin in 2002), it will be necessary to maintain a reserve force several times that number. Lacking the ability to test aging weapons, an alternate warhead design should be maintained for each of the major systems in the inventory. Allowances should be made for weapons in the refurbishment pipeline, those that are temporarily disassembled for maintenance, or those in transit to or from their duty station. Finally, some weapons with special capabilities should be kept "just in case."

Maximalists argue that it is foolish to dismantle thousands of weapons for no reason other than to say that we have done so. Our weapons represent a significant investment and should be maintained as a hedge against unforeseen contingencies. Other countries may or may not follow our lead in dismantlement.

Critics of the maximalist approach, myself included, note that it is wasteful to spread limited resources across thousands of weapons when only hundreds are required to deter aggression. The price tag associated with doing almost anything to a nuclear weapons system is at least a billion dollars, and that money would be better spent in improving the reliability and safety of a smaller set of weapons than in maintaining a huge arsenal in a warehouse.

The Moderate Position

I believe that the number of nuclear weapons we maintain should be based upon a careful analysis of military requirements with a hedge provided for unforeseen strategic contingencies. This I label the moderate position.

Fewer nuclear weapons will be required in the future because other means are available to achieve the same military objectives. Precision-delivered conventional weapons are capable of destroying many of the targets formerly assigned to nuclear weapons. Mobile missiles, submarines, aircraft, and many other types of targets do not *require* the explosive force of a nuclear weapon for their destruction. Nuclear weapons are required only for targets that would be very difficult or impossible to destroy with conventional explosives.

The exact number of weapons needed depends on our assessment of enemy assets and the level of confidence we would like to have in destroying them. Assuming that we maintain our counterforce strategy, we will most likely target only those enemy assets that could inflict grave damage on our country, our military forces, or our allies. These include nuclear and perhaps biological weapons, military command centers, and other locations of strategic significance.

Surveillance satellites can count large objects on the ground, including fixed missile silos and support facilities for nuclear submarines. We can thus count the number of things that we wish to hold at risk and size our force accordingly. As for the remainder of the adversary's weapons, including aircraft-mounted bombs and cruise missiles, they are typically concentrated in a small number of hardened locations, each of which could be destroyed with one or a few weapons. Any adversary with ballistic missile submarines will surely have at least half of its fleet at sea during a time of crisis—shifting the military requirement from nuclear weapons to attack submarines that can hunt down and destroy the enemy's missile carriers with conventional torpedoes. I took mobile missiles off the nuclear targeting list since they can be destroyed with conventional weapons assuming adequate intelligence.

The Russian Federation has by far the largest nuclear force of any potential adversary; it sets the requirement for numbers of weapons. Russia expects to have several hundred SS–27 missiles available in the near future and it will likely maintain some older ICBMs as a hedge. An American stockpile of five hundred to one thousand ready weapons would be sufficient to deter this force, that is, to make it impossible for Moscow to eliminate our weapons and avoid devastating retaliation following a first strike. Add to this an additional five hundred to one thousand weapons to cover the potential for system failure, the presence of weapons in the refurbishment pipeline, and some special weapons for unique applications, and the United States can still maintain a strong deterrent force with approximately one thousand to two thousand nuclear weapons—comparable to the number of weapons cited in the Moscow Treaty.

Having determined numbers, we can ask what *types* of weapons are required. A detailed analysis of potential targets shows that very few require more than ten kilotons for their destruction if the weapon is delivered with sufficient accuracy. Most of the remaining ones, all of which fall in the very hard target category, can be destroyed with a five-hundred-kiloton earth-penetrating weapon. If five hundred kilotons is not sufficient, it is unlikely that any explosion will suffice. A stockpile in which 90 percent of the weapons had ten kilotons of yield and the remaining 10 percent had five hundred kilotons is compatible with most future targeting requirements. Owing to the vulnerability of aircraft in reaching targets deep within enemy territory, all our weapons should be placed on ballistic missiles.

So far we have discussed weapons in terms of their explosive yield. Safety, security, reliability, and ease of manufacture are also critical to an effective stockpile. All the weapons in the current American nuclear arsenal were designed to maximize the yield-to-weight ratio, an economic consideration that enabled them to be carried on smaller missiles and aircraft. Since their designers anticipated that they would remain in the stockpile for no more than about twenty years, none were designed with long life in mind and none have a sufficient supply of spare parts. Only a handful were configured as earth-penetrating weapons that can hold very hard targets at risk. Finally, many have yields that are much greater than are required to satisfy future mission requirements. Our nuclear weapons stockpile is optimized for a threat that no longer exists.

Is there a role for manned bombers in the future of nuclear weapons? Perhaps, but only as a backup measure should an adversary develop an effective means of defeating ballistic missile warheads, a remote but nonetheless conceivable possibility.

Manned bombers have an advantage in that they can be recalled right up until the time that the weapon is dropped, but they suffer from relative slowness compared to ballistic missiles and vulnerability to anti-aircraft weapons and fighters. It can take many hours for a bomber to fly from the United States to a distant target, compared to the thirty to sixty minutes required for a ballistic missile to cover the same distance. If a manned bomber (one containing a pilot) is to be maintained, it should be the stealthy B2 rather than the venerable but vulnerable B52. Unmanned bombers (versions of existing unmanned aerial vehicles) could also be used, although the possibility that they might lose contact with their controller and wander off on their own is a serious worry that should limit their use for nuclear missions.

Cruise missiles launched from bombers or submarines lack both the flexibility of bombers and the assured delivery of ballistic missiles, thus rendering them least suited to nuclear missions of the future. Powered by jet engines, they are slower than ballistic missiles but nevertheless fly on their own without intervention of a pilot. The possibility that a cruise missile might be shot down is very real, and should this happen, its warhead could be captured by an adversary. This might be the most powerful argument for their removal from the arsenal.

AS STRANGE AS it may seem during a time when we are busy dismantling nuclear weapons, we need to build new ones. We need weapons with lower yields, and weapons that can destroy WMD threats buried in reinforced structures and under mountains. And we need weapons that have modern safety and security features, those that make the weapon nearly impossible to detonate if it were to fall into the wrong hands.

When designing the nuclear stockpile of the future, especially one that has a tenuous connection to tested designs, it is important to include *redundancy*. Two types of weapons are required—a ten-kiloton and a five-hundred-kiloton warhead—but it would be wise to field two designs for each to avoid the possibility of single point failure of the entire fleet. Should one type of weapon be found to have a serious problem, it could be withdrawn for repair or replacement, leaving the other design to carry out the deterrent function.

It is imperative that a viable manufacturing capability be established if we are to maintain any credible nuclear deterrent. Too many special processes are involved in the construction of a nuclear weapon to replicate them on demand, especially in a time of crisis. A regular program of remanufacture, like the ones currently being followed by all the other nuclear weapons states, will help maintain skills and equipment. Regular international inspections can assure other countries that we are replacing weapons rather than increasing our total stockpile.

THE FUNDAMENTAL ROLE of nuclear weapons in assuring the security of the United States did not change with the end of the Cold War. They still serve as the ultimate defenders of our freedom, weapons that deter any state-level adversary from attacking our vital interests. However, nuclear weapons are only part of our future strategic posture. Ballistic missile defenses, while limited at present, will be effective against nuclear attacks launched by North Korea, Iran, or other new entrants to the nuclear arena. Advanced conventional weapons can replace nuclear weapons in many missions, lessening our reliance on the latter while maintaining military capability. Improved transparency and inspection treaties with other countries would reduce the need to maintain nuclear forces larger than required and could conceivably enable us to eliminate them altogether.

Nuclear weapons have been, since their very inception, instruments of contradiction—built with the hope that they would never be used, the ultimate engines of destruction designed to maintain the peace. The United States has the ability to transform the strategic future, reducing the threat of nuclear war while defending ourselves in a dangerous world. If we continue on our present course, we will find ourselves lingering in the shadow of the Cold War, unequipped to face a century that will demand new solutions to ever more complex problems.

Index